BRISBANE BREACHED

The Story of a Drought Defaulted Floodplain

David Topp

Connor Court Publishing Pty Ltd

Published in 2024 by Connor Court Publishing Pty Ltd.

Connor Court Publishing Pty Ltd.
PO Box 7257
Redland Bay QLD 4165
sales@connorcourt.com
www.connorcourt.com

ISBN: 9781922815781

Cover Design by Maria Giordano

Printed in Australia.

Cover photos and photo credits: Kedron Brook, inner North Brisbane, on the western and eastern sides respectively of the Gilbert Road bridge, taken on 9 December 2019 during a demonstrable drought phase (David Topp); and on 27 February 2022 during a 'rain bomb' (Justin Fleming).

Endorsements

Some two centuries after the first Brisbane River explorers John Oxley and Edmund Lockyer reported seeing evidence of the past occurrence of significant floods, Barrister David Topp has comprehensively explored how the drought-prone City of Brisbane subsequently and regrettably came to be built on a flood plain. In the process he forensically records (in words, graphs and photographs) our city's tragic history of "over the floor" flooding, and frankly observes that the problem yet to be solved concerns "rain that does fall does not move on". *Brisbane Breached* is a disturbingly readable sequel to his *Tennyson Breach.*

(Tim O'Dwyer: Property lawyer, consumer advocate and author of *Real Estate Escapes*)

David Topp succinctly explains how a city built on a flood plain will never be flood free. He also argues that if Brisbane's citizens, businesses and governments don't do a better job of water conservation, the city could again experience severe water shortages. Brisbane puffery will be in overdrive in the lead up to the Olympics, but the issues raised by this book will not go away.

(Peter Spearritt: Emeritus Professor of History, The University of Queensland)

David Topp weaves the intricate nature of the different facets of politics, flood mitigation, water security and urban planning into an engaging story. It's one that helps explain the complex interconnections that plague flood plain management in a modern context.

(Rob Ayre, Hydrological and hydraulic engineer)

Contents

1

Drought: The Long Term Default

'*Rain rain go away and don't come back to Plainland ever again*'.

So sang my grandmother and her friends, then aged six, in 1910, frustrated by rain stymieing their hopes of playing outside.

Perhaps understandable given that the variety of indoor activities available to little children in those days was, to say the least, lacking compared to modern day offerings; iPads, YouTube, Spotify and TikTok were yet to grace the good people of Queensland of the early 1900's with their presence.

Some may opine that that was not necessarily a bad thing.

Anyhow the moral of that story, told to me some 80 years later, was be careful for what one wishes for, my grandmother concluding her vignette: '*The rain then went away, almost immediately, and for a number of years it seemed like rain would never return to Plainland ever again*'.

Plainland is located approximately halfway between Brisbane and Toowoomba. Notably it is, in longitudinal terms, close in location to Wivenhoe Dam, Brisbane's primary water supply reservoir. A dam which will be referenced frequently in this work.

Long term droughts akin to that my grandmother and her friends regretted having spawned by song are de rigueur for those environs: '*Despite being on the 'wet' east coast of Australia the dominating physical characteristic of the Brisbane River catchment is that it is essentially a 'dry' part of a dry continent. This 'dry' catchment has: a low average rainfall / runoff ratio; large volume floods; occasional severe droughts with zero stream flows; and, a pan evaporation equal to the average rainfall*'.[1] Then Justice of Appeal Catherine Holmes,

[1] '*The Brisbane River: A source-book for the future*', Eds Peter Davie, Errol Stock and Darryl Low Choy, The Australian Littoral Society, 1990, p. 3.

who helmed the 2011 Queensland Floods Commission of Inquiry (the "QFCI"), prefaced her report that '*there is no doubt*' that the flooding Her Honour was commissioned to inquire into '*took a state more accustomed to drought by surprise*'.[2]

As this graph of '*May – April Queensland area averaged rainfall for 1900 – 2010*' makes clear, the long term average line spends more time in the negative than in the positive[3]:

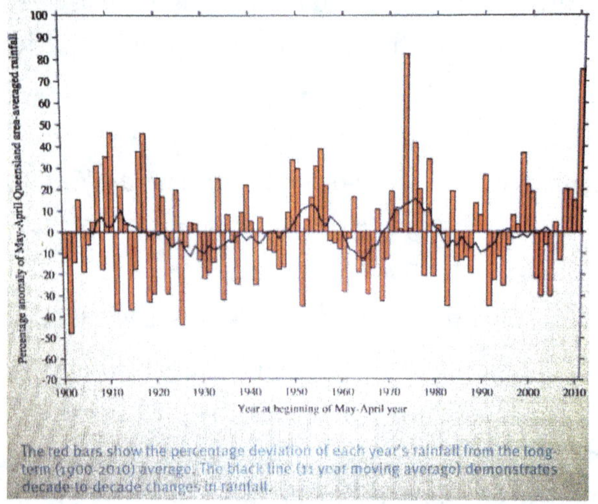

The red bars show the percentage deviation of each year's rainfall from the long-term (1900-2010) average. The black line (11 year moving average) demonstrates decade-to-decade changes in rainfall.

16 years after 1910, when my grandparents married on 26 April 1926, they endured another year blighted by drought, including a need to have sourced artificial flowers for the ceremony due to fresh versions being unavailable.

Proving that both in the days following imprudent childrens' songs in 1910 and for many years thereafter, long term dry spells, far from unfamiliar, are in fact the default in both South-East Queensland and Queensland generally: Bob Hawke's leading of the Australian Labor Party out of the post 1975 dismissal of the Whitlam Government wilderness was due in part to widespread despair, taken out on the incumbent coalition government of Malcolm Fraser, occasioned by a long term drought. A similar prolonged dry

---

[2] *QFCI Final Report*, p. 30.

[3] Queensland Department of Environment and Resource Management, Queensland Climate Change Centre of Excellence 'Queensland rainfall – past, present and future', December 2017, p. 5.

spell between 1991 and 1994 culminated in bushfires around this writer's Grade 12 final exam period in September 1994. The comparatively rainy second half of that decade in Brisbane was swiftly replaced by what became dubbed the Millennium Drought between 2000 and 2009, the longest drought since the Federation Drought at the commencement of the 20th century[4]: '*It was a decade so dry that as the trees withered from their outer branches we were all watering our lawns with buckets of water collected through our timed showers*'.[5]

The Millennium Drought broke in South-East Queensland gradually between June 2007 and 2009, with more substantial rains in 2010 enabling Wivenhoe Dam to reach its nominal capacity of 100% in October 2010.[6] It was a close-run thing though. Brisbane's existential future was imperilled by Wivenhoe Dam's capacity falling to a mere 15.08% in August 2007[7]; what was once a vast lake turned into an ugly dustbowl as land designed to be permanently inundated and hence wholly inarable was suddenly exposed to an unrelenting sun. What minuscule rain did fall was instantly lapped up by the parched dam floor. Only the gradual reinstatement of rains after August 2007, over the next three years, finally started to make Wivenhoe Dam look respectable again.

Until Wivenhoe Dam turned to look like something to be feared, not merely respected, when its capacity approached 188.7% in January 2011; the epic rain event of late 2010 and early 2011 created what is well known as the fourth of Brisbane River's major floods caused during genuinely reliably recorded history {1841, 1893, 1974 and 2011}.

High levels of consistent rain continued in and around Brisbane for three further years until another prolonged dry spell began after a rainy month of May in 2015 ended. Apart from a few heavy

---

[4] Stuart Layt, 'Cloudy with a chance of hyperbole', *The Saturday Age*, 9 April 2022, p. 31.

[5] Editorial, 'With rain on way don't forget 2011', *The Courier-Mail*, 11 October 2022, p. 30.

[6] Ibid.

[7] Current Dam Levels, *The Brisbane Times*, 17 August 2007.

rain breaks in March 2017 and January and February 2020, this just over six year dry spell did not let up until, in typical Queensland style, roof destroying hailstorms at Springfield Lakes, south-west of Brisbane, on 31 October 2020. It was therefore state election day in Queensland which, entirely coincidentally, heralded a restoration to a more consistent phase of rain, albeit in unspectacular numbers totals wise.

This Bureau of Meteorology graph vividly shows the juxtaposition between the extreme levels of Wivenhoe Dam, at both ends of the spectrum, in one single decade: from a high of 188.7% at the height of the 2011 flood event to a very worryingly low 36.3%. Bearing in mind of course that after a prolonged drought phase, rains which finally start falling again over dams and in their source tributaries are absorbed by parched ground before being able to convey waters to storage reservoirs as intended[8]:

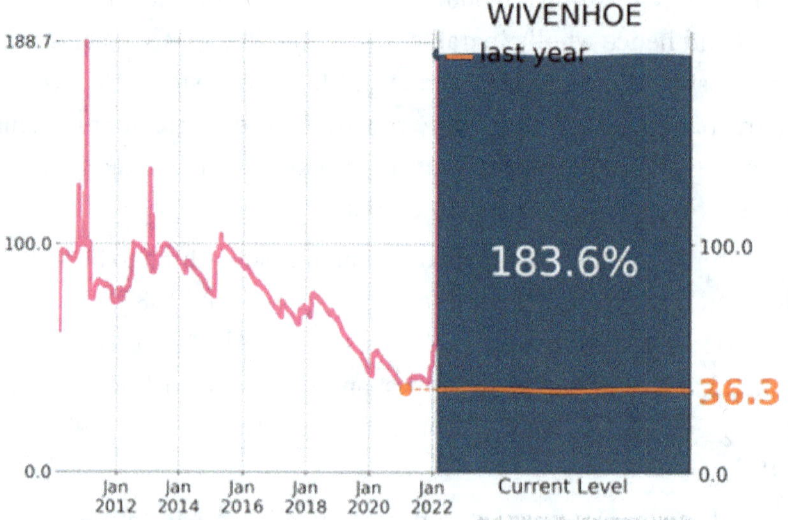

My grandmother was not ultimately able to live to see an end to the millennium drought, she having passed away in 2005. Having seen many dry-spells during her long life, no rain events were, whether by words spoken or sung, exhorted by my grandmother to

[8] Bureau of Meteorology, 'Special Climate Statement 76 – Extreme rainfall and flooding in south-eastern Queensland and eastern New South Wales', 25 May 2022, p. 15.

ever again '*go away*' during her further 95 years lived after the song she regretted that her six year old self ever chose to sing. The reasoning being obvious: Brisbane, and South-East Queensland generally, is a perennially drought prone environment, drought being the long term default.

As exemplified by the demonstrable downward trend in Wivenhoe Dam levels between mid 2015 and January 2021 depicted by the Bureau's graph. Similar long term falls in that Dam's levels also occurred during the other two broad drought phases cited in this chapter in the 16 years between 1991 and 2007.

Rain, when it does fall in Brisbane and the broader South-East Queensland region during these many such phases, is therefore universally welcomed.

Unless, of course, the rain that does fall does not move on.

As occurred during the final week of February 2022.

2

The '*Rain Bomb*'

Chapter 1 essayed a long term phase of low rainfalls in and around Brisbane between 2015 and election day on 31 October 2020.

Gradually moderate rains then ensued during the 15 months to the final week of February 2022.

Then the bomb hit.

The bomb?

The '*unrelenting rain bomb*' of February 2022 as it was dubbed.[9]

Purists of meteorology may have bemoaned the crassness and coarseness of the term: '*Every meteorologist I know hates this term. I have no idea where it came from*' lamented Queensland University of Technology oceanographer and climate researcher Mark Gibbs.[10]

However it was anything but inapt to describe the sheer enormity of the rains that fell.

'[T]*his rain bomb is just unrelenting. It's coming down in buckets – it's not a waterfall, it's like waves of water just coming down*'[11] was the way Queensland's Premier at the time Annastacia Palaszczuk described, on 27 February 2022, what she and Brisbane's population could not avoid encountering on that extraordinary day.

Mt Glorious, the name given to both a 599m high mountain north-west of the Brisbane CBD and the suburb constructed upon it, is a day-trip and bushwalking location known and enjoyed by visitors to and residents of Brisbane alike.

---

9 Kate Kyriacou, 'Rising waters for next three days', *The Courier-Mail*, 28 February 2022, p. 2.

10 Stuart Layt, 'Cloudy with a chance of hyperbole', *The Saturday Age*, 9 April 2022, p. 30.

11 Kate Kyriacou, 'Rising waters for next three days', *The Courier-Mail*, 28 February 2022, p. 3.

Not that bushwalking or pleasant barbeques were possible during the final four days of February 2022 when 1776mm – 1.776 metres – of rain fell on Mt Glorious and nearby environs. Neatly encapsulated and described by consultant meteorologist Anthony Cornelius of Weatherwatch this way: '*1776mm in 4 days, nearly as much as 1974 AND 2011 together*'[12]:

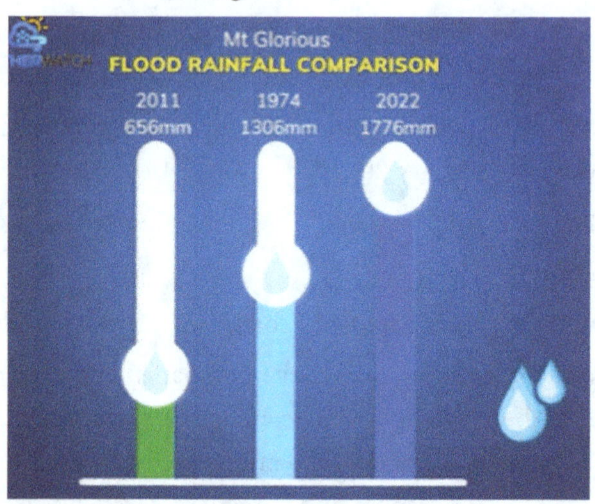

Mt Glorious being the peak of the falls, other South-East Queensland totals included[13]:

Alderley	1007mm.
Redcliffe	967mm.
Brisbane	792.8mm, 676.8mm of which – which in turn equates to 80% of Brisbane's annual rainfall[14] – fell in just three days.[15]
Brisbane Airport	719mm.
Karalee	695mm.
Caloundra	616mm.
Gold Coast Seaway	332mm.

[12] Anthony Cornelius Meteorologist, Facebook 1 March 2022.

[13] Jack Mckay and Thomas Chamberlin, 'Fast and furious to slow and very painful', *The Courier-Mail*, 1 March 2022, p. 3.

[14] Sarah Richards, 'A break down of how south-east Queensland's flood crisis played out', *ABC News Online*, 21 March 2022.

[15] Brisbane City Council, 'Living in Brisbane', March 2022, Special Edition, 'Brisbane unites after floods'.

Brisbane's 792.8mm was an all time record, exceeding the previous record received contemporaneously with the 1974 flood of 655.8mm.[16] 'Rain of Terror' was the pun-intended cover headline of the 26 February 2022 *Courier-Mail*, 'Flood of Tears' being the next day cover headline of the 27 February 2022 *Sunday Mail.*

The Bureau of Meteorology concluded that for large areas of South-East Queensland, the final week of February 2022 was the wettest week since at least 1900.[17]

Needless to say, comparisons with the major Brisbane flood event of 2011, a mere 11 years earlier, were quickly made. '*It's just like '11 all over again*' being one such newspaper headline.[18] Other commentators asked: '*How could we be here again, so soon after the horror of the one-in-100-year flood of just 11 years ago? So much was the same as 2011 – the macabre spectator sport of watching boats still tethered to their pontoons hurtling down the muddy, swollen Brisbane River, the Rosalie shopping district under water, the road closures, the pain*'.[19]

The '*just like '11 all over again*' headline was very apt to have described the eerie similarities between 10 January 2011 and 25 February 2022 in Grantham, a small Lockyer Valley hamlet which garnered worldwide notoriety for all of the wrong reasons. During the afternoon and evening of 10 January 2011, severe flash flooding occurred in Toowoomba, the waters of which literally tumbled down the Great Dividing Range from Toowoomba's 600 metres above sea level location into the Lockyer Valley. Grantham was devastated by the torrent; houses were literally ripped by the surging waters off their stumps and floated away. Up to eight residents were rendered missing, feared dead and ultimately presumed dead as a result.

---

[16] Jack Mckay and Thomas Chamberlin, 'Fast and furious to slow and very painful', *The Courier-Mail*, 1 March 2022, p. 3.

[17] Bureau of Meteorology, 'Special Climate Statement 76 – Extreme rainfall and flooding in south-eastern Queensland and eastern New South Wales', 25 May 2022, p. 3.

[18] Jeremy Pierce, Felicity Ripper, Tom Gillespie and Michael Nolan, 'It's like '11 all over again', *The Courier-Mail*, 26 February 2022, p. 4.

[19] Leisa Scott, 'The brown snake has turned on us again', *The Courier-Mail*, 1 March 2022, p. 6.

A veritable carbon-copy event occurred on 25 February 2022; Grantham was yet again '*cut off in a frightening reminder of the carnage that swept the region 11 years ago..... Streets became ferocious river rapids as floating debris smashed into power poles, trees and parked cars....The Grantham township looks like a lake just appeared out of nowhere*'.[20]

The only thankful aspect being the loss of lives of Grantham residents caused in 2011 was not repeated in 2022, with property losses also being less severe.

Yet despite all of the disproportionate damage wrought to such a tiny town of Grantham on 10 January 2011, because none of those waters of devastation sourced from South-East Queensland's two primary dams – Somerset and Wivenhoe – in fact not from any dams at all – none of its residents could seek any recourse in the subsequent class action lawsuit analysed in detail at Chapter 5 of this work. As that action's trial judge, Justice Beech-Jones held, Grantham's '*flooding is not the subject matter of this case. There is no relevant connection between the conduct of any of the defendants* [Wivenhoe Dam owners and operating entities] *and the occurrence of that flooding. Instead, the large increase in flows in Lockyer Creek that occurred on that day is simply part of the factual background to the "over the floor" flooding that was occasioned at other places in the Brisbane River catchment on 11 and 12 January 2011 (and beyond)*'.[21]

Justice Beech-Jones was remarkably, albeit unintentionally, prognostic on the 29 November 2019 date of delivery of his verdict by invoking a notion of '*over the floor*' flooding. Because exactly that occurred in a vast number of places within and around Brisbane during February 2022: '*The combination of river, creek, and overland*

---

[20] Jeremy Pierce, Felicity Ripper, Tom Gillespie and Michael Nolan, 'It's like '11 all over again', *The Courier-Mail*, 26 February 2022, pp. 4 & 5.

[21] *Rodriguez & Sons Pty Ltd v Queensland Bulk Water Supply Authority trading as Seqwater (No 22)* [2019] NSWSC 1657, judgment summary, at [3]. From here on in, in both these footnotes and the text, this decision will be cited as '*First instance*', this term reflecting the fact that the NSW Supreme Court's trial decision was the first step in what proved to be a three stage hierarchical process of court determinations of what was commonly known as the 2011 Floodapps Class Action lawsuit.

flow flooding left significant damage, with more than 23,200 properties affected across 129 suburbs'.[22]

This is because, and unlike the obvious similarities between 2011 and 2022 in Grantham and suburbs in Brisbane directly affected by the Brisbane River, a remarkable amount of facts were anything but the same as 2011 in many other suburbs and regions of South-East Queensland. The broader city of Brisbane, especially its northern suburbs, fared relatively well in 2011, as exemplified in these four separate reflections:

- '*The Brisbane flood of 2011 happened on a bright and sunny day, the culmination of weeks and weeks of rain in far flung catchments*'[23];
- '*The pattern of* [January 2011] *rainfall caused little, if any, significant creek flooding in Brisbane, though creeks were flooded by backwater from the* [Brisbane] *river...The rainfall...caused little, if any, flash flooding in Brisbane's suburban creeks. The flooding that did occur in Brisbane's creeks was essentially due to backing up from the flood waters in the Brisbane River*'[24];
- '*The locations where the rainfall fell most heavily also differed between 2011 and 2022. In 2011 the heaviest falls were north and west of Brisbane. In 2022, they were more coastally focused south of Gympie including greater Brisbane....for Brisbane and Ipswich, flash flooding was more significant along smaller creeks and tributaries during the February 2022 event compared to January 2011*'[25]; and
- '*It was a river flood event, which meant that as the Brisbane river burst its banks...other suburbs were high and dry. As I drove between affected suburbs in my ward, I passed through*

22 Lord Mayor Adrian Schrinner, 'A Helping Hand', *Style Magazine*, April 2022, p. 52,

23 Leisa Scott, 'The brown snake has turned on us again', *The Courier-Mail*, 1 March 2022, p. 6.

24 Brisbane City Council '*Brisbane Flood January 2011 Independent Review of Brisbane City Council's Response*', pp. 1 & 13.

25 Office of the Inspector-General Emergency Management, 'South-East Queensland Rainfall and Flooding February to March 2022 Review', 31 August 2022, p. 34 of 155.

those that were on higher ground and unaffected...it was like a surreal parallel world. Residents in these suburbs were mowing their lawns, going for a jog. Some ... were even ringing ... to complain their bins weren't being collected (flooded roads were blocking access for garbage trucks). Yet just five minutes down the road, residents were in the process of losing all their worldly possessions'.[26]

Therefore so long as suburbs were neither adjacent nor near to the Brisbane River, they remained relatively unaffected by what was heavy but not extraordinary rainfall. Rather the previously cited '*far flung*' areas quite some way away from Brisbane central, like the Brisbane Valley {not to be confused with the inner city suburb of Fortitude Valley}, including Wivenhoe Dam and its vast catchment area, suffered the burden of massive amounts of rain.

2022 changed that script markedly; not only was it the all time record-breaker in terms of the sheer volume of the water that fell during a single flood event period, the vast and broad range of areas which came to be deluged meant that '*over the floor*' flooding emanated from many more sources than the Brisbane River itself.

'*This time it happened over four or five days...rain that just did not stop, day or night. Sodden ground could hold no more; local creeks that did not flood in 2011 began to fill and swell before creeping over the banks. It'll stop soon, won't it?*'[27]

It did stop, around 10pm on Sunday 27 February. The damage had more than been done by then of course. Two prime examples of the '*local creeks that did not flood in 2011 began to fill and swell before creeping over the banks*' phenomenon were these: Firstly, Kedron Brook, which is not a tributary at all of the Brisbane River. Secondly, Enoggera Creek; though it does in fact merge with the Brisbane River, its junction point is relatively close to the Brisbane

---

[26] Julian Simmonds, '*Yes Lord Mayor: Inside the Politics of Brisbane City Hall*', Connor Court Publishing, 2021, p. 83.

[27] Leisa Scott, 'The brown snake has turned on us again', *The Courier-Mail*, 1 March 2022, p. 6.

River's mouth, meaning that the backflow flooding that so adversely affected feeder systems upstream like the Bremer River and Oxley Creek during the 2011 flood event did not deleteriously impact suburbs on the Enoggera Creek banks that year.

As an aside, and for geographical naming context, Enoggera Creek actually changes name by the time of its intersection with the Brisbane River; at Windsor its name changes to Breakfast Creek. Perhaps fortuitously for the well-known 'Brekky Creek' historic pub on its banks at Albion; the slogan 'a steak and a beer off the wood at the Enoggera Creek' just doesn't have the same ring to it.

Naming anomalies aside, the upshot is that the majority of northern Brisbane suburbs suffered minimal loss or damage in 2011 as neither extraordinary rains fell over them and the Kedron Brook and Enoggera/Breakfast Creek systems coped with what rain did fall.

Not in 2022 however; Kedron Brook and Enoggera/Breakfast Creeks became raging torrents as unprecedented rainfall totals in their respective sources[28] fed into them. Both the Enoggera Creek and Kedron Brook systems' sources are quite close to each other: Mt Nebo for Enoggera Creek and Upper Kedron, 10 kilometres away, for Kedron Brook. This is significant because of both sources' relative proximity to Mt Glorious, mentioned at the commencement of this chapter due to its extraordinarily large quantum of rainfall. Those waters had to release somehow and released they duly were along whatever creek systems existed; Kedron Brook and Enoggera Creek, being the largest of the systems east and south-east of Mt Glorious, were inundated, meaning that any and all of those systems' natural release valves, being low lying banks, breached, leading to horrendous flash flooding of homes and businesses in suburbs like Toombul, Ashgrove, Wilston, Windsor and Albion.

Compounding the waters rushing eastwards from a saturated Mt Glorious were the vast amounts of rain which fell in Brisbane itself, especially over its northern suburbs. At 12.30pm on Sunday 27 Feb-

---

[28] Jeremy Pierce and Danielle Buckley, 'Mount Glorious has now copped more than 1m of water in less than two days', *The Sunday Mail*, 27 February 2022, p. 2.

ruary 2022, a date already cited in this book for its extremity, Anthony Cornelius of Weatherwatch captioned the below image this way: '*...a heavy line of showers and storms has developed along the secondary convergence zone. Falls of up to 117mm have been recorded in a single hour*'.[29] Note the red colouring, depicting the greater intensity of the rainfalls – east and south-east of Mt Glorious, which directly corresponds with the courses taken by Kedron Brook and Enoggera Creek:

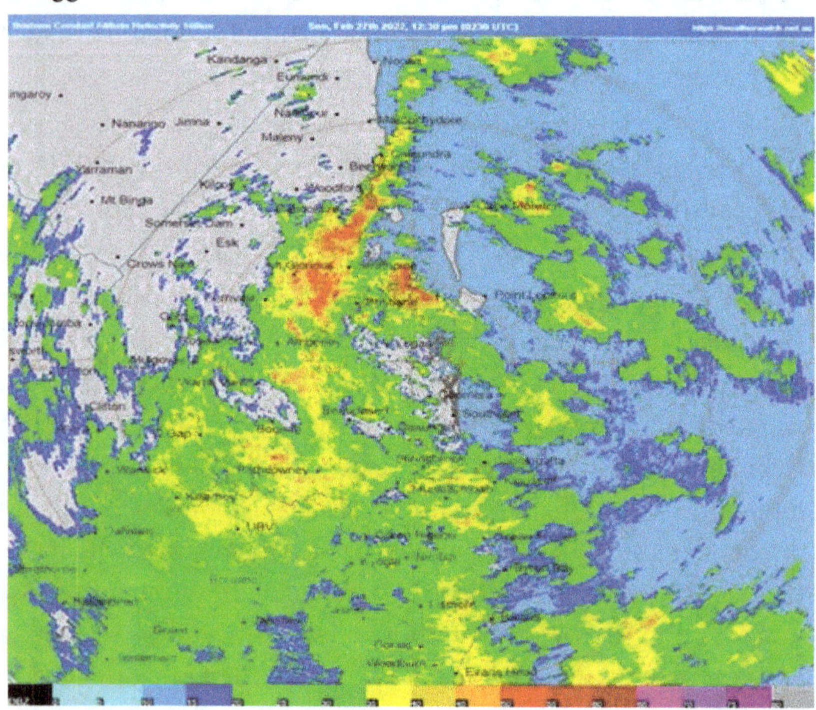

In Brisbane's inner northern suburb of Grange that 27 February 2022 day, local resident and weather watcher Hector Chesty's home monitoring station recorded Grange as having raised its bat to denote a century, not of test match runs, but rather of rain, as early into the day as 6.19am. The double century was reached at 10.29am, just over four hours later. 300mm was surpassed at 1.29pm, three hours later, then 400mm by 3.39pm, being two hours and 10 minutes later. It is therefore obvious that as the day progressed the 100mm totals

[29] Anthony Cornelius Meteorologist, Facebook, 27 February 2022.

were being received in lesser and lesser time intervals, due to the increasing intensity of the falls. By 11.59pm, Chesty's total count for 27 February was 552.50mm, more than half a metre of rain in one single 24 hour period.

The enormity of the 27 February 2022 rainfalls, especially when considering the previous day's 245.59mm and in addition the 263.09mm recorded by Chesty at Grange on 25 February 2022, caused 'backing up' of storm waters unable to naturally drain into the hopelessly overloaded Enoggera Creek and Kedron Brook systems which bypass Grange to its respective south and north. In turn causing backflow flooding quite some way away from their banks. By way of example, Grange's southern neighbouring suburb of Wilston possesses a long stormwater drain which conveys storm waters from its eponymously named Wilston Hill, under Edmondstone Street, and into Enoggera Creek. This self-taken picture depicts that drain's commencement of overflow on 27 January 2013, when the rain-depression remnants of the then ex-Tropical Cyclone Oswald caused commensurate heavy rainfalls in Brisbane and South-East Queensland generally:

As that particular day progressed, so did the overflows, causing the Edmondstone Street storm water drain to become wholly submerged:

Precisely identical effects on that particular drain were occasioned on the third day of the 2022 crisis, causing adjacent houses to be inundated to the point of unlivability. This self-taken picture depicts the march up the hill along Edmondstone Street in a westerly direction from the drainage canal in the direction of Newmarket's Finsbury Street, causing flooding to this house worse at the rear than at the front coextensive with its upper sloping land pitch, but rendering it unliveable either way:

Not exactly the best of sales pitches for the agency that had already been tasked with marketing, that's for sure.

Following those waters' recession, a close-up of that particular's property front door is this:

Whereas the back yard of that house, being the same garage visible at the much more inundated rear of house depicted in the street view picture, was almost completely consumed by the substantially greater quantum there of the floodwaters; note the much higher upper limit of the waters denoted by the respective strips of mud:

Across the road from the flood devastated house featured formerly, on the northern side of Edmondstone Street, is, or was, the renowned Crust & Company Artisan sourdough bread and croissants bakery and café. A Wilston institution, receiving patronage from all over the city for nine years, Crust & Company chose to announce during June 2022 that it would not rebuild.[30] Though no attribution of blame was made by the operators, the inherent risk of flooding along that street due to Wilston's long stormwater drain being unable to divert its waters other than into Enoggera Creek, with backflow upstreet inundation the inevitable result when the latter is itself overflowing, suggests they had no other choice:

As this overhead picture, though more than 50 years old, nonetheless is still capable of vividly depicting[31]:

PLATE 4.7 - Flood affected areas abutting parkland to the nor of the creek downstream of Normanby Bridge.

[30] Crust & Company Artisan bakery Facebook post 20 June 2022.

[31] Cameron, McNamara & Partners Pty Ltd, 'Report on Flooding in the Catchment of Breakdast and Enoggera Creeks', March 1973, at plate 4.7.

The flood-affected mud-striped house depicted earlier in this chapter sits under the stem of the ‘h’, the building housing Crust & Company sits across the roadway bisecting the middle of the photograph {Edmondstone Street}, Enoggera Creek is the curved waterway near the ‘a’ and the ‘e’, and the drainage channel can be seen emerging from the small forest where the ‘h’ is captioned extending in a broadly straight line to Enoggera Creek. Rugby purists might notice under the ‘e’ the Ballymore rugby union ground, which is much more substantial these days.

The long main road depicted to the left of the picture is and remains Newmarket Road. Between it and the letter ‘c’ today, viewing the image left to right, are, respectively, the Homezone homewares bulk shopping precinct, and a network of hockey and netball fields in what is thankfully still retained green space. In February 2022 floodwaters similarly escaped as far northwards from Enoggera Creek as Newmarket Road, which naturally dips as it approaches its intersection with Silvester and Noble Streets, causing ponding so pronounced that the dipping in the road was fully concealed. These pictures are of that particular intersection looking, respectively, from the north, east and west:

Officeworks

While this image depicts Homezone having become flood zone, causing anguish and heartbreak for the many small business traders in that complex, some of whom were yet to re-open as late as December 2022:

Turning to the broadly parallel to Enoggera Creek in a path sense Kedron Brook, it transits, amongst other inner northern Brisbane suburbs, Alderley, which was referred to at the beginning of this chapter as having received a weekly total of 1007mm, and this writer's home suburb, adjacent to Alderley, of Grange. All of that water similarly had only one outlet.

Kedron Brook is also well known to many Brisbane residents for the adjacent Kedron Brook bikeway which meanders along its banks. Used by both commuter and recreational cyclists alike, along with many thousands more walkers and runners, its long held popularity spiked even further during the height of the Corona Virus induced lockdowns of 2020. Local Brisbane City Councillor Andrew Wines observed that during this time more people used the Kedron Brook bikeway than were using the Brisbane CBD's Queen Street and associated Queen Street Mall,[32] so denuded of patronage by the exhortations of governments for shops to close, staff to work from home and people generally to quit the city in the name of suppression of the virus. As a nearby resident and father of two children desperate to escape the cabin fever of life at home, bearing in mind the official

[32] Councillor Andrew Wines, ABC Radio *Brisbane Drive* interview, 23 March 2022.

exercise exemption to lockdown, I can attest that Cr Wines was correct. People were literally everywhere out and about on the Brook.

During 2017 the Brisbane City Council completed a project to light the Kedron Brook bikeway for safety's sake, given the number of persons it attracts and the obvious risks during the winter months when later sunrises and earlier sunsets darkened the bikeway during peak times of day for commuter cyclists and before and after work recreational runners and walkers.

Plus the versatility the lighting – assisted this time by a rising moon in the east – gave to time-poor local parents whose kids needed pre-bedtime energy burnt off by outside exercising!

Lindsay Millard illustrated {in red circling} on this image of his the way in which those very useful light poles bent and snapped in the direction of the downward flows of Kedron Brook on 26 and 27 February 2022, never to be capable of illuminating the path ever again:

The cost involved in replacing the lighting infrastructure forms part of the Brisbane City Council's damages bill of up to $330 million, three times more than the cost of the 2011 flood, as a result of the full extent of Council assets affected by the February 2022 event being a list as eye-watering as this: 198 buildings on community-leased facilities, 106 sport fields, 74 car parks, 285 kilometres of roads, three bikeways, seven pontoons, 4282 street and bikeway lights, 863 parks, 300 playgrounds, 40 public barbecues, 25 park toilet blocks, 22 public pools and 77 parking meters.[33] By comparison, and proving the vastly greater extent of overall loss and damage to Brisbane in 2022 compared to 2011, damage to Brisbane City Council assets and facilities caused by the 2011 flood cost the Council $106 million adjusted for inflation, which sum was '*blown out of the water*'[34] by this year's drenching.[35]

Also blown out of the water were hopes of many for two new 'green bridges' – that is pedestrian and cyclist only – over the Brisbane River at two West End locations. A revitalisation of East Brisbane's Mowbray Park and restoration of the Council's monohull ferries were yet more examples of projects also announced as shelved by Lord Mayor Adrian Schrinner as a result of the flood damage clean up and repair spend.[36]

Compounding the difficulties for Schrinner's Council – this time at the other {incoming revenues} end of the budgetary equation – more than 12,000 ratepaying property owners affected by the flooding applied for a flood rebate. In addition, 800 owners formally objected to their unimproved land valuations on a flooding pretext,[37] reason being that the higher those values the greater the amount of rates local authorities can charge and the state government can levy

[33] Cloe Read, 'Flood forces Brisbane council to pause projects for 'biggest-ever rebuild'', *Brisbane Times* 19 May 2022.

[34] Pun may or may not have been intended.

[35] James Hall, 'February flooding bill worst on record', *The Courier-Mail*, 12 May 2022, p. 11.

[36] Hayden Johnson, 'Council projects go down the river', *The Courier-Mail*, 20 May 2022, p. 7.

[37] Jack McKay, 'Flood of valuation fears', *The Courier-Mail*, 10 June 2022, p. 10.

on account of land tax, hence the obvious incentive for those owners to talk down their values.

That the Gilbert Road overbridge which links Lutwyche to Gordon Park in the background to Lindsay Millard's fallen light poles picture is as high in excess of Kedron Brook's water level as depicted would surely guarantee it to be high and dry no matter how epic a flood event could be? How could it not, especially when considering how wide the Brook's banks are to absorb any overflows? February 2022 however rendered those presumptions perilously close to incorrect. Justin Fleming's image of the Gilbert Road bridge on 27 February 2022 is the epitome of the proverb that a picture tells a thousand words. Actually, Fleming's well exceeds that; it tells more like 1 million words {with a few expletives undoubtedly thrown in also}:

And yes that is a shipping container wedged firmly against one of the pillars, the container having been wrested away by the forces of the waters and conveyed until eventually finding a resting place of sorts.

Fleming's images say just about all. The only thing they can not explain is the terror of what those raging waters looked and sounded like in real time. I stood on the nearby Damon Road, some 80 metres downstream, myself that 27 February 2022 afternoon to take the following picture. In normal times when Kedron Brook represents the small quiet stream that its benign name suggests, one can not see the waters from Damon Road as they are located too far down in the gully to be viewed from the road reserve. Not that the waters were hard to spot on 27 February; my image depicts the extremely rapid rise in the Brook that day (and also, in the background, the same Gilbert Road over bridge photographed by Fleming):

For perspective of the rapid rise, note the submerging of most of the shared pedestrian/bike path sign just to the left of the main pole holding up the playground shade screen. The playground is actually high up on a ridge; the bike path drops down flush with the Brook's reserve, with the water level – usually – being lower still. Another juxtapose is comprised by the following set of three self-taken photographs, two of which were taken from the Gilbert Road bridge itself, which did not actually go under, though it was an extremely close run thing; as Fleming's picture makes clear, there was precious little daylight between it and the peak of the waters. Note the small pergola type covered seat at top of the first image on 27 February which had the raging flood waters lapping at its foundations compared to how high that structure sits atop Kedron Brook in normal times; from the pergola itself, being the third picture in the set, one has to consciously look downwards to actually view the natural water level, thereby proving how extra-ordinarily high the volume of water was that last Sunday in February:

Meanwhile downstream of Grange, where Kedron Brook transits its eponymous suburb of Kedron, this was the situation at Shaw Road's number one rugby field: the tops only of the goalpost pads were visible. Due to the small amount of daylight between the top of the Kedron Brook overflow flood waters and the cross-bar, only the tallest of that club's second-row forwards would have been able to have kept breathing!

The only aspect all of the images in this chapter can not depict is the rage aspect; one just had to be there to fully comprehend, not that it would have taken long, that it would have been a case of instant sweep-under and carriage to a certain death had any person been foolish enough to have set foot in that poisonous brew of rainwater, mud and myriad other detritus that simply never stood a chance of remaining in situ once Kedron Brook turned violent, and then came to be wrapped around whatever remained standing as the worst of the waters, finally, receded:

The only reason why the white caps in the first of Fleming's photos are not all that white at all is because there can be no white water when the water itself is so stained by mud, and not just mud flowing naturally into as mere run-off, but mud sourced by the Brook in fact stealing its own banks. This picture depicts actual land having been literally carved out by Kedron Brook upstream at Mitchelton, adjacent to the Brookside Shopping Centre:

Thankfully Brookside Shopping Centre was both high enough and far enough away from Kedron Brook to have escaped any damage. Unlike its counterpart Toombul Shopping Centre downstream; the entire Toombul Shopping Centre located on Sandgate Road along Kedron Brook's bank was inundated and destroyed to such an extent that its owner, Mirvac, predicted a full six months of closure necessary to clean up and rehabilitate the centre to a standard fit for reopening.

A question that would however beg from a concept of not merely landslides but landstrips is how would mere houses near the banks of the Brook have coped?

These pictures tell the story of the flood height firstly and then the sinister under the surface carve-outs and traumas caused by the torrent which soon came to be glaringly evident: note firstly the white painted house on the northern bank of what was the wholly sodden Kedron Brook in the suburb of Gordon Park on the afternoon of 27 February:

Note also the cars in the background of this zoomed in version of the above picture. That is not a causeway those cars are traversing; rather that's the road bridge linking the suburbs of Gordon Park and Lutwyche, a bridge high enough to incorporate the Kedron Brook bikeway below. All of that height was fully consumed by the torrent:

Now compare these post-flood pictures taken late in April 2022, almost two months later. Note how vulnerably that same white coloured Gordon Park house stood after the flood waters receded, having removed the neighbouring park's hillside. The tarpaulins weighed down by rocks being necessary to ensure that future rainfalls drained directly away rather than contributing to further undermining of the already impaired land structure:

Barely 500 metres upstream yet another Kedron Brook adjacent home owner was required to take the same evasive action; hopefully the foundations of that particular house are sitting more firmly than the precarious looking fence!

In Gordon Park's neighbouring suburb of Grange, on the southern bank of Kedron Brook, not just land but concrete bikeways were literally severed and dispensed, presumably as far away as into Moreton Bay itself. Well known Brisbane graphic artist Tony Bela released his photo superimposed by labels to describe the way in which Kedron Brook so radically altered between the end of February 2022 and the commencement of March 2022:

Optus 4G 10:32 am 73%

Tony Bela · 12 m ·

I put this together to show a small part of the destruction this 2022 flood has caused, every suburb in Brisbane has a story to tell, this is just one of the many from our suburban home.

The two photographs below are best viewed in conjunction with the first of the two cover photos, as they depict what little is left of the narrower of the two parallel paths on which my children were riding their scooters during a very dry day during late 2019. Severed by the forces of the waters. The original bank of the Brook was where the three very flimsily grounded trees depicted on the overhead view photograph are. It is clear that more than double the width of the Brook has been reclaimed by it as a result of the forces of 27 February 2022 waters:

Further upstream, this time in the suburb of Stafford, the same raw brutality is evident from these photos taken by Alan Sagan:

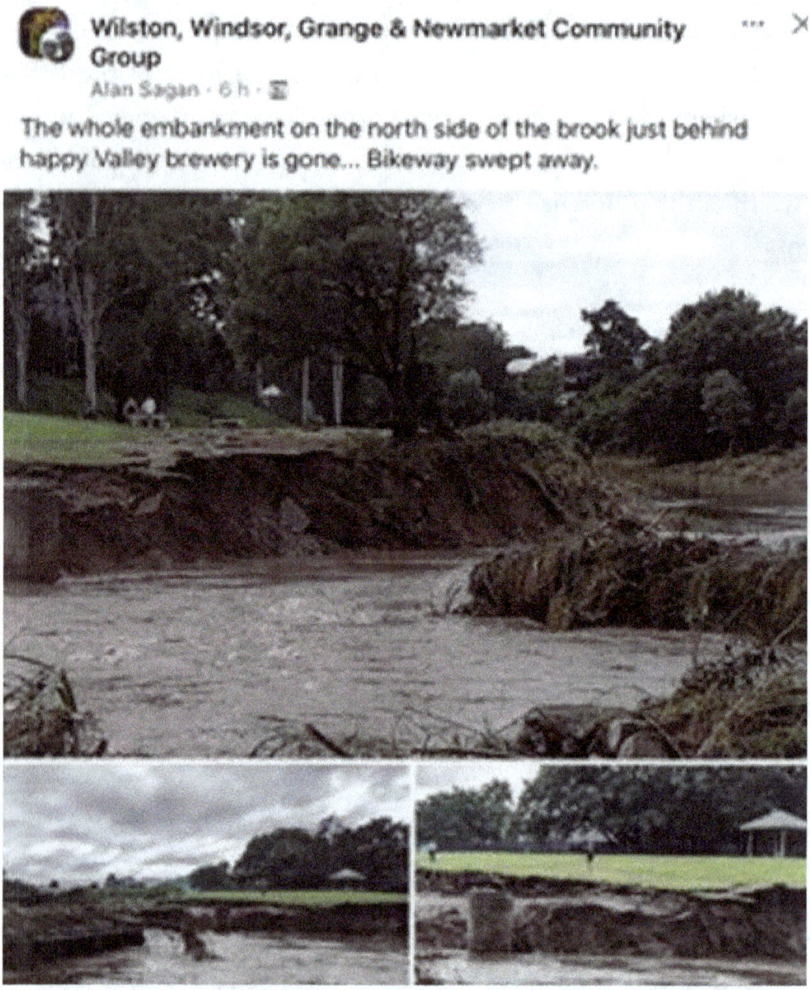

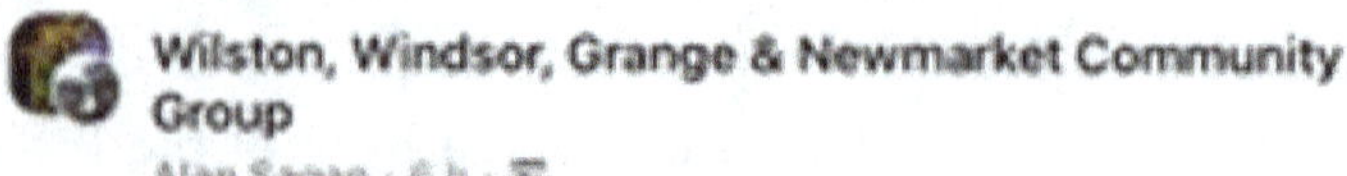

Wilston, Windsor, Grange & Newmarket Community Group
Alan Sagan · 6 h ·
The whole embankment on the north side of the brook just behind happy Valley brewery is gone... Bikeway swept away.

I returned, some 100m upstream from Sagan's vantage point, two months later to photograph what happened just to the left of Sagan's pictures: the Kedron Brook Bikeway causeway bridge which connects Stafford in the east and Grange Forest Park in the west was severed from its connecting land embankment when the embankment was swept away, leaving a gap as yawning as depicted in this picture:

Eleven months later the Kedron Brook Bikeway causeway bridge remained just as well and truly cut off as my April 2022 picture depicted. Reason being that the same flood also damaged an adjacent sewer main, itself requiring repair before filling and reinstating the embankment and reconnection of the bridge could occur.[38] Based on these self-taken 14 February 2023 photographs, the just under 1 calendar year delay is explicable given the extensive rock reinforcing necessary to prevent a repeat the next time the Brook rages:

[38] Andrew Wines Brisbane City Councillor for Enoggera, 'Out and about in the Enoggera Ward', *The Everton Echo*, July 2022, p. 21.

To put into perspective the sheer power of the Brook on 27 February 2022, please compare these two pictures which first appeared in your present writer's post-2011 flood event analysis book *Tennyson Breach.*[39] I took these photos at Cardwell, a Bruce Highway township located roughly halfway between Townsville and Cairns, which I visited five months after the Category 5 Severe Tropical Cyclone Yasi menaced an area of coastline stretching from Cairns in the north to Bowen in the south. Cardwell bore the brunt of the system; the devastation of the early days of February 2011 was still raw and evident in June 2011. Including the way in which a concrete jetty fractured; the images are of the same jetty both on it and from the side to better depict the split:

[39] David Topp, '*Tennyson Breach*', Boolarong Press, 2012 ("Tennyson Breach"), pp. 106 & 107.

That the Kedron Brook bikeway causeway bridge could also be rendered as useless in the same way as the Cardwell jetty was proves that the powers and forces of a monster Category 5 cyclone of 2011 were carbon-copied 11 years later and some 1500 kilometres to the south. It is submitted that the force of suburban Brisbane floodwater on 27 February 2022 genuinely equalled the destructive effect of a Category 5 cyclone.

Symbolic then that, in the same way that it is traditional for the Premier of a cyclone-prone state such as Queensland to visit areas damaged by serious cyclones, Kedron Brook's damage garnered an in-person inspection from both the Premier and Deputy Premier:

Jimmy Sullivan MP

MEMBER FOR STAFFORD

COMMUNITY NEWS | AUTUMN 2022

The impact of the severe weather. Joining Premier Annastacia Palaszczuk an State Recovery Coordinator Major General Jake Ellwood to see the extensive dam caused by the recent floods along the Kedron Brook.

IN THIS EDITIO

FLOOD RECOVERY

We're working to assist our community to get back on their feet following the impact of the recent severe weather.

FRONTLINE SERVICES

The Palaszczuk Government is investing record funding into our hospitals and adding more personnel to our frontline.

COMMUNIT BENEFIT F

More than $354,000 fo community groups, P& sporting clubs in grant up to $35,000 each.

The name given in the QFCI Report to the resulting damage to river and creek banks depicted in the many Kedron Brook photographs published in this chapter, and similarly caused by other extreme rainfall events elsewhere, is 'land slumping':

> *... the Commission received several submissions from people whose property had suffered severe erosion or bank slumping (where chunks of riverbank become unstable and topple or slide into the river) during or immediately following the January 2011 flood event at Wivenhoe Dam. These submissions provided a stark reminder of the magnitude of the flood and the impact the river can have on land bordering it.*[40]

One of which submissions depicted the loss of approximately eight hectares of land from a farm downstream of Wivenhoe Dam[41]:

That waters spilling not naturally but artificially as a result of manual releases from a reservoir as vast as Wivenhoe Dam could cause such an extent of loss and damage is nonetheless explicable; less so is that none of the land slumping along Kedron Brook was the result of any artificial flows like intentional releases from a previously closed gated dam causing a consequential surge of water. To the contrary,

[40] QFCI Report, p. 574.

[41] QFCI Report, p. 575.

those vast and powerful flows along Kedron Brook on 27 February 2022 were entirely natural; Kedron Brook is wholly undammed. Indeed, the damage so sustained can, in hindsight, be explained as a natural and inevitable consequence of the extreme intensity of the rainfalls recorded at the Brook's cognate suburb of Grange that day when 552.50mm was received, one 100mm increment of which fell in as little as two hours and 10 minutes. Six months after the loss and damage to Kedron Brook was finally assessed and quantified, the rebuild bill for its eponymous bikeway was predicted at $50 million over two years,[42] proving the truly extraordinary power of the raging floodwaters of 27 February 2022.

A total of 19 deaths and at least two others remaining missing was the toll to 10 March 2022 of the summer of 2021-22 for South-East Queensland.[43] The largest flood caused loss of life since 33 perished in the infamous summer of 2010-11. Given the ways in which even concrete was no match for the forces of the February 2022 floodwaters in Kedron Brook, the death toll was not disproportionate, sad as of course any loss of life is. If ever the mantra invented by the Queensland Government to dissuade people from driving through floodwaters of 'If it's flooded, forget it' was singularly appropriate, it was during February 2022. The comparatively low death toll for what was an undeniably all of Brisbane rain and flooding event, and an extremely serious one at that, suggests that the message finally made it through.

Lastly let it not be forgotten that wealthier nations can absorb natural disasters in ways which minimise losses of life so much more than in places less well off; during April 2022, areas of Durban, in South Africa, endured up to 450mm of rain in a 48 hour period, causing a total of 341 fatalities.[44] Yet the extremity of Durban's event was less than Brisbane's; Durban's 225mm per 24 hour

---

[42] Andrew Wines, Brisbane City Councillor for Enoggera, 'Budget 2022-23 Projects Commence In Our Local Area', *The Everton Echo*, September 2022, p. 24.

[43] Danielle Buckley, 'Faces of our victims', *The Courier-Mail*, 10 March 2022, p. 10.

[44] AFP, 'Durban pelted with half its yearly rain in 48 hours', *The Weekend Australian*, 16-17 April 2022, p. 8.

average equalled Brisbane's 676.8mm over 72 hours,[45] namely 225.6mm. The difference being three consecutive days rather than 2, which was an all time record for Brisbane as recorded by the Bureau of Meteorology:

> *There were three consecutive days, 26 to 28 February, with totals over 200 mm; across the current and former sites there have been only eight previous days with totals over 200 mm, none of them consecutive. When compared across the current and former sites, the Brisbane City gauge set records for all periods from 3 to 7 days.*[46]

The upshot being that both Brisbane and Durban endured identical rain intensity, though despite Brisbane's being received over three consecutive days rather than 2, and consequentially 226.8mm more in total, Brisbane's fatality rate was a tiny 5.57% of Durban's.

Proving how disproportionately natural disasters harm and hurt the less well off.

---

[45] Brisbane City Council, 'Living in Brisbane', March 2022, Special Edition: 'Brisbane unites after floods'.

[46] Bureau of Meteorology, 'Special Climate Statement 76 – Extreme rainfall and flooding in south-eastern Queensland and eastern New South Wales', 25 May 2022, p. 12.

3

Brisbane is a Flood Plain

Chapter 1's thesis was that Brisbane is inherently drought prone.

Whereas Chapter 2 cited the 2011 and 2022 flood events as proving the myriad ways in which Brisbane is inherently at risk of severe flooding, and suffers accordingly.

Contradictory much?

Politicians who attempt vastly contradictory positions and policies according to the daily tenor of debate are inevitably held, often with vehemence, to account by media and their political opponents. Credibility being a necessary asset for all political aspirants and participants to possess, a charge of hypocrisy is often the death-knell to the ever-necessary credibility politicians desire.

The Brisbane River however runs for no office. It answers to no-one. It seeks no permissions. Indeed attempts by governments past and present to 'tame' the Brisbane River have proven spectacularly unsuccessful.

A River With a City Problem ('River') is the thought-provoking title to Margaret Cook's history of the symbiotic relationship between the City of Brisbane and its eponymous river.

Published in 2019, Dr Cook was remarkably prescient in concluding her work by decrying '*a well-travelled path: reliance on Wivenhoe Dam underpinned by an ongoing belief that floods can be tamed with a technocratic solution*' and opining that '[F]*uture floods are inevitable*'.[47]

Not even three years passed before the inevitability Dr Cook's work predicted transpired.

[47] University of Qld Press 2019, pp. 193 & 197.

Another writer was identically vindicated, he having forecast 11 years before 2022 that '[F]*loods of equal or even greater severity to January 2011 will affect Brisbane again*'.[48] And though the 3.85 metre Brisbane River at the City Gauge flood peak on 28 February 2022 was lower, numerically, than the cognate figure in 2011, the proposition of Chapter 2 is that the loss and damage wrought to Brisbane in 2022 was of a greater severity due to the vastly inflated number of suburbs which suffered compared to those during 2011.

Brisbane is not known as 'the river city' for nothing, given the way in which the Brisbane River bisects the city from Moggill in the south-west, through the Brisbane Central Business District in the centre and out to its mouth at Pinkenba in the north-east. Flooding of the Brisbane River therefore becomes extremely serious for suburbs on its banks which are flush relative to the river itself, such as Jindalee and the eastern halves of Graceville and Chelmer, or suburbs 'fed' by river overflows along intersecting creek systems, such as Goodna, Rocklea, Auchenflower, Rosalie and Milton. Milton of course being the home of Suncorp Stadium, an iconic stadium for rectangle shaped football codes, beloved by all Queenslanders and viscerally hated by those south of the border following 40 years of bitter rivalry in State of Origin. Save only for the citizens of Penrith and associated western Sydney suburbs following 2021's special one-off only National Rugby League grand final held outside of Sydney due to NSW Government imposed corona virus lockdowns when the pandemic affected Sydney more so in 2021 than during 2020. Suncorp Stadium was definitely not built for swimming however. Let alone if the water covering the hallowed turf is not chlorinated but rather an ugly shade of flooded brown.

Milton's Suncorp Stadium quickly recovered to host a first round National Rugby League 2022 season fixture on 11 March 2022. However Milton's state primary school was a write off for the rest of term 1; it and five other flood damaged South-East Queensland school

---

[48] *Tennyson Breach*, p. 131.

campuses did not re-open for lessons until Tuesday 19 April 2022, after a total recovery spend by Education Queensland of $50m.[49]

Conventional wisdom has long held that Brisbane suffered two monster floods in its history: 1893 and 1974. At height levels at Brisbane's City Gauge[50] of 8.35m and 5.45m respectively, the epithet is not unwarranted. 2011, at 4.46m formed the triumvirate and only a mere 11 years later the highest height recorded at the City Gauge was 3.85m on 28 February 2022.

Interestingly though, these four events may not be the full extent; Cook's historical researches in River added that '[A]*lthough records are scarce*', evidence exists for a flood with an 8.43m flood height in 1841[51]; if so, it would have exceeded even 1893's level. Further back as far as Captain James Cook's voyage along the eastern seaboard in 1770, on board botanist Joseph Banks is said to have journaled a '*dirty clay colour*'[52] in the waters of what is now known as Moreton Bay, Moreton Bay being the name given to the body of water that separates greater Brisbane in the west from North and South Stradbroke and Moreton Islands in the east. Press photos taken of the beautiful blue Moreton Bay, known to most Brisbanites as a classic location for boating and fishing, instead turning into an ugly oil-slick looking brown as the 2011 floodwaters finally receded to sea, made for anything but pleasant viewing. Proving both that the sole end-point for Brisbane River floods is Moreton Bay, and that flood waters are inevitably brown as a function of all the mud, dirt and soil which overland waters scrape up and remove when floods occur. If Joseph Banks' reported observation of May 1770 was correct, the inference capable of arising is that major flooding of the Brisbane River could have occurred no less than six times in 252 years.

Which translates to an average of 1 in every 42 years.

---

[49] *ABC News*, Queensland telecast, 17 April 2022.

[50] 'City Gauge' being the title of the Brisbane River flood level measuring station in Brisbane's Central Business District at the corner of Edward and Alice Streets.

[51] River, p. 10.

[52] River, p. 3.

Markedly less than the 1 in 100 year flooding that politicians were quick to allege after 2011; this book's predecessor, *Tennyson Breach*, analysed the way in which the 2011 floods devastated the Tennyson Reach complex of apartments at the inner southern Brisbane suburb of Tennyson. Causing residents to have been rendered homeless for at least three months as their beleaguered apartment buildings were cleaned out and restored. Nor were lot owners given rent assistance for their hurriedly found alternative accommodation; not only did those hapless residents have to self-fund both their mortgages and their newly imposed emergency rentals,[53] to add insult to injury, all were invoiced $10,000 special body corporate levies per lot to fund the complex's clean up and rehabilitation works.[54]

In so doing I challenged the notion, put during this post-2011 flood quotation from then Lord Mayor of Brisbane Campbell Newman, whose 2004 – 2011 term in that office coincided with the development approval of and subsequent construction to completion of the Tennyson Reach complex:

> *The Tennyson* [Reach] *development ... is approved to deal with a one-in-100 year flood, plus another 500mm. That's the standard we work to and that's a pretty rigorous standard. What happened out there is extremely severe and rare flooding. I reckon it's a lot more than a one-in-100 year event.*[55]

The January 2011 flood level at Tennyson Reach was confirmed by surveyors said to have been retained by its developer, Mirvac, at 9.05 metres.[56] In 1974, in the site's power station days, the level was 10.8 metres,[57] and in 1893 the estimated flood level was 13.65 metres.[58]

---

53 *Tennyson Breach*, p. 91.

54 *Tennyson Breach*, pp. 92 & 93.

55 Milanda Rout, 'Developer to help clean up luxury complex', *The Australian*, 20 January 2011, p. 6.

56 Mirvac circular letter to Tennyson Reach residents, 20 January 2011, third un-numbered paragraph.

57 Mirvac circular letter to Tennyson Reach residents, 20 January 2011, fourth un-numbered paragraph.

58 Estimate taken from David A Dunworth's QFCI submission {first un-numbered page}.

Meaning that, far from *'a lot more than a one-in-100 year event'*, the flooding that afflicted Tennyson Reach in January 2011 was, on an average of those statistics and years, a far more mundane 1 in 39.66 year event.

Trying therefore to explain away Brisbane River flood events with epithets like one in 100 years is futile. And if that wasn't apparent post-2011, the fact that the city flooded yet again a mere 11 years later ought finally terminate such rhetoric. The verifiable truth is that the Brisbane River floodplain is Australia's most flood impacted area,[59] vividly captured by this diagram which juxtaposes the floodplain area with the catchment as a whole[60]:

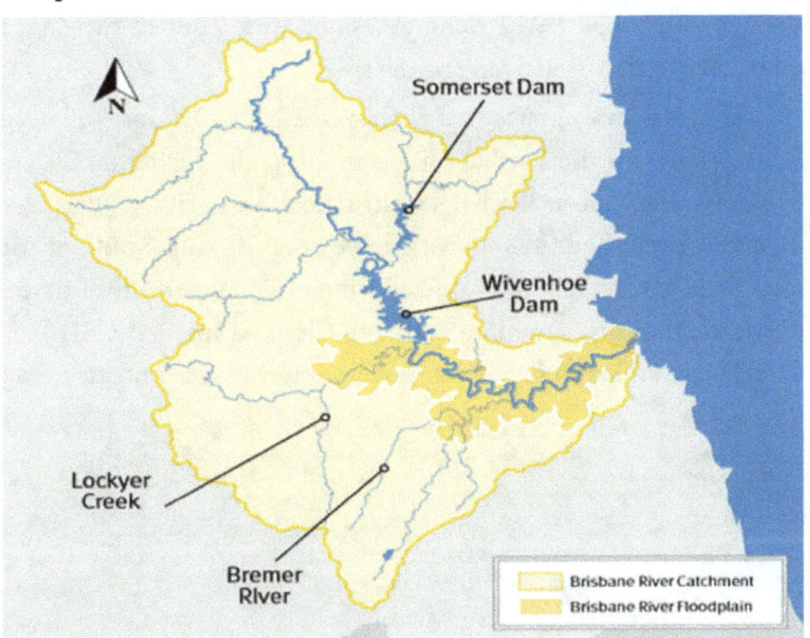

Compounding the difficulties of the Brisbane River floodplain being the nation's largest are the facts that an estimated 134,000 buildings were estimated, in 2017, to be located on that very flood plain – itself thought to be the largest number of buildings on any

[59] *'Brisbane River Strategic Floodplain Management Plan'*, a joint initiative of the Australian Government, Queensland Government, Brisbane, Ipswich, Lockyer Valley and Somerset Councils and Seqwater © The State of Qld {Qld Reconstruction Authority} February 2019, p. 8 {"The Plan"}.

[60] The Plan, p. 14.

floodplain nationwide[61] – and 280,000 people resident on the floodplain {as of 2018}.[62] In the event of a 1% {1 in 100} AEP flood,[63] about 17,300 of those 134,000 buildings would experience flooding, two thirds of which are located in the Brisbane City Council's jurisdictional area.[64]

Turning now to that concept of 'AEP', flood risks are not to be described and interpreted in literal one in 100 year terms. Rather the correct terminology is Annual Exceedance Probability {AEP}. AEP refers to the probability of a nominated flood height occurring in any given year. Therefore a 1% AEP flood describes a flood event with a 1 in 100 chance of being equalled or exceeded in any given year, whereas a 50% AEP denotes a flood with a one in two chance of occurring across a given 365 days.[65]

There is a 55% chance that at least one 1% AEP flood will occur at least once to the Brisbane River flood plain during an 80 year window. The Queensland Reconstruction Authority {'QRA'} very helpfully published the following table of AEP's which not only depict their percentage likelihoods during notional periods of time of 80 years, the corresponding Brisbane City and Ipswich CBD flood heights are added to provide necessary interpretive context[66]:

AEP	At least once in 80 years	At least twice in 80 years	Brisbane City gauge (m, AHD)	Ipswich CBD (m, AHD)
10% (1 in 10)	100%	100%	1.8	14.8
5% (1 in 20)	98%	91%	2.2	16.1
2% (1 in 50)	80%	48%	3.2	18.7
1% (1 in 100)	55%	19%	4.5	20.1
0.2% (1 in 500)	15%	1%	7.3	23.4
0.05% (1 in 2000)	4%	0.1%	9.9	25.7
0.001% (1 in 100,000)	0.1%	<0.1%	23.7	36.1

[61] The Plan, p. 25.

[62] The Plan, p. 26.

[63] See paragraph below for explanation of 'AEP'.

[64] The Plan, p. 25.

[65] The Plan, p. 20.

[66] The Plan, p. 21.

Whereas there is only a 4% chance that, during a period of 80 years, one single 1 in 2000 AEP rainfall count would be received. Readers of Chapter 2 will recall the way in which local Grange resident Hector Chesty recorded a total rainfall depth of 552.50mm in a 24 hour period during the February 2022 event (with the peak burst occurring on 27 February). Comparing this 24 hour peak rainfall depth to Brisbane City's Intensity Frequency Duration curves demonstrates that, for the suburb of Grange that very day, the 4% chance during the notional 80 years actually occurred:

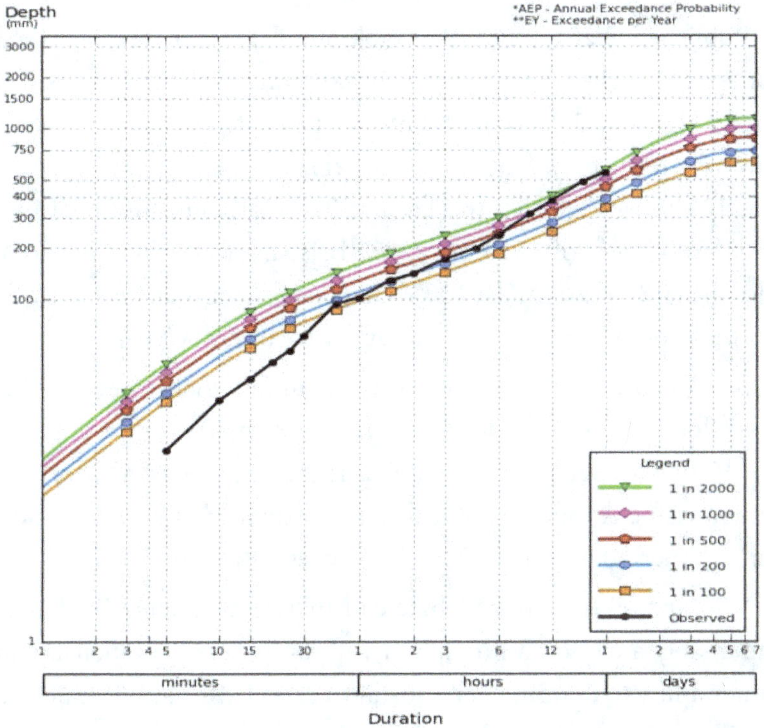

{Graph credit: Steven Voss, Surface Water Engineer}

Again proving the genuine extraordinariness of 27 February 2022, a day that will live on in Brisbane's history in perpetuity.

Meanwhile the following QRA diagram demonstrates how the concept of AEP's corresponded to the actual flood levels observed during the 1893, 1974 and 2011 events[67]:

[67] *The Plan*, p. 22.

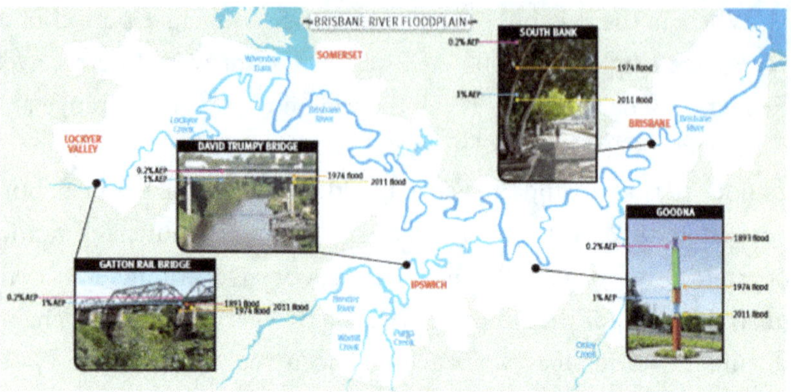

Published in 2019, the QRA's table and diagram necessarily could not apply the Brisbane River's February and May 2022 City Gauge flood heights of 3.85 and 1.80 metres respectively.[68] Doing so now however, retrospectively, classifies February 2022's 3.85 metres as exactly halfway between the 1% and 2% AEP's, meaning that there is a 67.5% chance of a February 2022 Brisbane River flood occurring once every 80 years, and a 100% chance of the minor May 2022 follow up flood occurring not once but twice every 80 years.

As of course happened, there were no less than three Brisbane River floods registering flood heights of 3.85 metres or greater at the City Gauge between 1974 and 2022. Three in 48 years: an average of 1 in 16 years. A much greater frequency than what a 67.5% chance during a period as long as 80 years suggests.

What possible further debate can therefore be had? Brisbane is undeniably a city on a floodplain with an inherent propensity and vulnerability to flooding of a substantial and loss-making scale, and, as will be analysed at paragraph 7 below, a flood plain facing an increasing vulnerability.

Where debate will however continue to ensue is in respect of the topic of dams.

---

[68] For more on the May 2022 minor Brisbane River flood, see Chapter 9, '*May Mayhem*' below.

4

Lies, Dammed Lies and Statistics

Benjamin Disaraeli has been quoted as inventing the well-worn phrase about there being three categories of lies: Lies, Damned Lies and Statistics.

Exactly what the difference between lies and damned lies is was not explained by Disraeli, nor seemingly can anyone else shed the necessary light.

Dammed lies however are easier to particularise and identify – the lies that pertain to dams, both as regards the science of dams and also the politics of dams.

Of that there is no shortage. Especially in this state. Only in Queensland would a premier, on the night of his resignation after a 19 year premiership, look back and declare: '*Clive, I've built a lot of dams*'.[69]

That premier was Sir Joh Bjelke-Petersen.

And the '*Clive*' was Clive Palmer, a then staffer of Bjelke-Petersen, nowadays better known for his mining wealth and political party Palmer United Party which became the United Australia Party, with their ubiquitous yellow and black liveries that dominated advertising for the 2013, 2019 and 2022 federal elections.

Palmer's recall of Bjelke-Petersen's whimsical reflection on his dam-building prowess was disclosed during a televised debate for the 2022 Federal Election featuring himself, Bob Katter, Pauline Hanson and, in an interesting development following the end of Chapter 3, Campbell Newman. Newman, after relinquishing the Brisbane Lord Mayoralty, won the 2012 Queensland State Election,

---

[69] Clive Palmer, speaking in the *Sky News* 2022, 'Mavericks' pub test debate', broadcast on 20 April 2022; Alice Workman, 'Campaign Diary', *The Australian*, 22 April 2022.

lost his premiership three years later and tilted at a Senate seat for the Liberal Democrats at the 2022 Federal Election. The *Sky News* debate covered numerous topics, though, being of national focus, flooding in Brisbane in 2011 and 2022 was not one of them.

Dams are reflexively proffered by many as the silver bullet capable of flood-proofing towns and cities along river banks. Especially in Brisbane given its history. Or perhaps histories: not only of major Brisbane River floods but also, as theorised in Chapter 1, droughts:

> *South-east Queensland is not situated in a generous, reliable rainfall belt…in the summer months sunshine is of sufficient severity to evaporate significant volumes of stored water'*, causing Brisbane to have *'had water shortages in the 1940s and throughout the 1950s.*[70]

This historical record is supported by reporting during 1952 that, despite a week's worth of rains that boosted Somerset Dam's storage to 7000 million gallons, then Lord Mayor Sir John Chandler still felt compelled to order '*that water restrictions would be continued*'.[71]

Something clearly had to have been done:

- '*After extensive investigations … into water storage in south-east Queensland and a feasibility study completed in June 1971, construction of Wivenhoe Dam began in 1977*'[72]; and
- '*It was the official view that Somerset Dam would be sufficient to supply Brisbane's needs until 1980. However, a dissenting view, expressed in 1950, that Brisbane would need additional water storage proved correct … in 1971 the State Government recommended a site … at Wivenhoe, well suited to the dual purposes of water storage and flood mitigation and where construction would be less expensive*'.[73]

---

[70] Denver Beanland, *Brisbane: Australia's New World City*, Boolarong Press, 2016, p. 205.

[71] *The Courier-Mail*'s 70 years ago extract from the 22 March 1952 edition, extracted p. 35 of the 22 March 2022 edition.

[72] Beanland, op. cit., p. 206.

[73] Helen Gregory, *The Brisbane River Story – Meanders through Time*, Australian Marine Conservation Society Inc, 1996, p. 83.

Interesting timing in that the 1974 flood bisected the two years cited in the first of those two quotes: 1971 and 1977.

It is therefore contended that the largest of Brisbane's dams – Wivenhoe, which is frequently referenced in this work – was not conceived following and solely as a reactive result of the 1974 flood event.

Nor the 1893 flood event either; even the extremely high water volumes conveyed during February 1893 had, during the ensuing 77 years, been and gone under the metaphorical bridge by then: '*Three generations had passed since the flood of 1893, and many people had become complacent about floods*'.[74] The contemporaneity of this quote – having been published in 1976 – is apposite.

Also a limited hydro-electric generation capacity existed as a part-pretext to Wivenhoe's creation; the Splityard Creek 500 megawatt hydro-electric power station in the background of this picture still part contributes to broader South-East Queensland electricity generation today, operating on a basis of waters pumped upwards from Wivenhoe during off-peak hours to enable deployment of hydro generation during peak hours:

[74] J. G. Steele, *The Brisbane River – an illustrated history*, Rigby Publishers, 1976, p. 53.

Hence why, rather than flood proofing, a need for greater water supplies was the primary pretext for Wivenhoe's creation. Given that Wivenhoe Dam is three times the capacity of its upstream feeder dam, Somerset Dam, clearly Wivenhoe Dam was and remains of supreme importance to prevent an entire city running out of water.

A scenario that looked increasingly likely to transpire during the dark days of 2007 when Wivenhoe Dam's capacity fell to 15.08%.

Additionally there was one other dam proposal also cited during the early 1970's: a dam at Wolffdene, south of Beenleigh, some 50 kilometres south of Brisbane's CBD. However, after the 1974 floods, '*construction of the Wolffdene Dam had been passed over in favour of … Wivenhoe… because the latter could be utilised for flood mitigation purposes*'.[75] This was because Wolffdene would have dammed the Albert, not Brisbane, River, both rivers being some many kilometres apart and wholly independent of each other; the Albert River does not contribute in any way to flooding within the Brisbane City Council's jurisdictional area.

The 1974 flood therefore having made the construction of Wivenhoe Dam a fait accompli, work began and Wivenhoe Dam formally opened in 1985. Premier Bjelke-Petersen, at the official opening, was quick to doubt that Brisbane would ever flood again. Other commentators at the time took a similar view:

> *What is being done to prevent another disaster like the 1974 flood? … flood mitigation works will be continued, including the erection of the Wivenhoe Dam … after this dam has been built floods like those of 1893 and 1974 will occur less frequently*'.[76]

So broadly entrenched had the view become that the approval, then finalisation, of Wivenhoe Dam would permanently and undeniably floodproof Brisbane, the actual opening and Premier Bjelke-Petersen's comments were barely even treated as news. A minuscule report of Wivenhoe Dam's official opening, not even by-lined by a

---

[75] Denver Beanland, *Brisbane: Australia's New World City*, Boolarong Press 2016, p. 205.

[76] J. G. Steele, *The Brisbane River – an illustrated history*, Rigby Publishers, 1976, p. 56.

journalist, appeared in a blink and you will miss it segment of page 18 of the then broadsheet *Courier-Mail*'s 19 October 1985 edition, sharing identical column inches with an adjacent story of a Melbourne prison governor summonsed to appear in the Heidelberg Magistrates Court to answer 53 criminal charges stemming from his alleged mismanagement of the Fairlea womens' prison. So blasé had the populace apparently become that such a nothing story of zero relevance whatsoever to Queensland shared equal column billing to the news headline of '*Flood threat past: Sir Joh*'.

As for what little the paper did report, it is instructive to note Premier Bjelke-Petersen was quoted as blithely writing off the resumption of 350 properties with historical homes having to be demolished as part of the permanent flooding process to have created Wivenhoe: '*But that's what happens with progress*'. That Premier Bjelke-Petersen won a thumping majority in his National Party's own right at the next year's state election, not even needing traditional coalition partner the Liberals to command a majority in the newly elected parliament, infers that those 350 property owners similarly accepted that theirs was a sacrifice that ought to have been made so as to guarantee future flood-proofing of Brisbane, and either omitted to protest or did so so quietly that no difference at all was made to the result of the 1986 state poll.

Another prominent politician of the era, Brisbane's 1985-1991 Lord Mayor Sallyanne Atkinson, also rode the change in mindset by having '*designated 1987 as the Year of the River to focus attention on its recreational potential ... It was a major exercise in community involvement, with groups working on riverfront parkland and developers being made aware of* [Brisbane] *river views*'.[77] 1987 was not only 13 years after the 1974 floods, providing genuine temporal separation from the unpleasantness, but was one single year before what was to become the highly successful World Expo 88. Held in 1988, it was, as its name suggests, a six month long international exposition staged at what is now Brisbane's South Bank precinct. Organisers

[77] Sallyanne Atkinson, *No Job for a Woman*, University of Qld Press, 2016, p. 170.

deliberately attempted to leverage its riverside location to boost the event's appeal; having attended Expo myself, as an 11-year-old, I can recall that they succeeded in that particular endeavour. And overall also; the city of Brisbane truly 'grew up' as a result.

The net cumulative result of the success of World Expo 88, the commissioning of Wivenhoe Dam, its associated perceived flood immunity and the 1987 Year of the River initiative, was that by the time the 15th anniversary of the January 1974 flood event was marked in early 1989, the Brisbane River had well and truly rehabilitated its reputation amongst the population.

The Brisbane River did at least wait until six years after Sir Joh Bjelke-Petersen's 2005 passing away to commence its course of proving his remarks at the opening ceremony wrong. No attribution of dammed, or damned, lie ought therefore be made against a deceased person, especially where what he said was a prediction, predictions inherently at risk of not being fulfilled. Rather the gravamen of the pun intended chapter title is that dams will mean many different things to many different people and politicians and accordingly give rise to claim and counterclaim and many more besides, with no ultimate resolution being possible.

By way of example of the intractable conflicts, take this comment piece published but 1 week after the worst of the February 2022 rainfalls ended:

> *We could build bigger and better dams which would require the expenditure of considerable political and fiscal capital and the resolve and strength to stare down the Greens who would howl like wolves at the mere mention of these play-things of the Devil.*[78]

Though the latter concept is obviously meant to be a debating point, the sheer cost of fiscal capital O'Connor cited is a statement of proven fact, not contestable opinion. '[A] *whopping $5.4 billion*

---

[78] Mike O'Connor, 'Time to drop a rain bomb on this insanity', *The Courier-Mail*, 8 March 2022, p. 61.

to build the long talked about Hells Gate Dam in North Queensland … dwarfs almost any announcement made during the [2022 Federal Election] *campaign*'.[79] Albeit subject to a '*business case*' not expected to be delivered until June 2022. Notwithstanding that rather important caveat, the project was nonetheless heavily promoted by the incumbent Morrison Government in the hope that the hefty fiscal capital outlay would reciprocally generate a commensurate sum of political capital for the then upcoming 21 May 2022 federal election.

As per usual, there were many varied views quoted in *ABC News*' report. Local vegetable growers lauded the potential for better water security. Whereas other irrigators were concerned about the potential for Hells Gate's waters to be the most expensive water in Australia.

Another notable point was the irregularity of Commonwealth money funding the project, and, in so doing, disabusing irrigators of their stated concerns. The following quote was attributed to Barnaby Joyce:

> *Usually state governments develop water projects and they're supposed to charge irrigators enough to cover the costs, but the Deputy Prime Minister has hinted that this water could end up being a "gift" to Queensland farmers from federal taxpayers. "We are not asking for a return".*[80]

Other than a return from the voters at the 19 May 2022 Federal Election of course.

Be the topography and water supply requisites of North Queensland as they may, it is difficult to conceive of how or where future dams in South-East Queensland could be built so as to attempt to hold back the 48.416 percent of the Brisbane River inflows which are sourced from sub-catchments downstream and hence wholly independent of Wivenhoe Dam.

Firstly, the Lockyer Valley, which comprises the Lockyer Creek

---

[79] *ABC News* Queensland telecast, 5 May 2022.

[80] Ibid.

sub-catchment of the Brisbane River, is not known as a food bowl for nothing; vast quantities of fruit and vegetables are farmed there for consumption both at home and abroad. Damming such fertile and productive farmland is untenable.

Secondly the Bremer River sub-catchment of the Brisbane River is rapidly being developed for housing as the inexorable growth of the South-East Queensland population continues apace. 'Dawn Walloon', a new housing estate 15 minutes west of Ipswich, is an example of one such estate, accumulating attention from first home owners wishing to take advantage of the $230,500.00 average entry price for vacant land due to the suburb of Walloon representing the second most affordable new land location available in South-East Queensland.[81] And not just for locals but also '*buyers from south of the border chasing ... value and ... Queensland sunshine and lifestyle*',[82] thereby increasing the region's collective attractiveness for new home buyers nation wide.

A study into potential new dam sites commissioned and published during the three-year Campbell Newman premiership was quite bereft of ideas for suitable new dam locations, notwithstanding Newman's urging that '[I]*t was a proper engineering study by the officers of the department*'.[83] In December 2014, the study so commissioned by Newman's government considered a list of nine potential dam sites in all, yet could only settle on two which warranted more detailed studies.

The first of which was on Warrill Creek, south-west of Ipswich, and, crucially, a feeder tributary to the inherently flood prone Bremer River.[84]

So on a flood mitigation pretext alone the proposal may have made sense. It did not, however, proceed any further: *The Warrill*

---

[81] Sophie Foster '*Land sales are losing the plot*', The Sunday Mail, 30 October 2022, p. 31

[82] Ibid.

[83] Tony Moore, 'Dam or be damned: Campbell Newman revisits 2014 plan for flood protection', *The Brisbane Times*, 12 May 2022.

[84] Lydia Lynch, 'Deluge defence report shelved', *The Australian*, 10 March 2022, p. 5.

Creek dam option was considered in 2021 but ruled out'.[85] Though no reasons were attributed, this writer submits two objective propositions for why a Warrill Creek dam could not and can not proceed: Firstly conflict due to the way in which the Warrill Creek catchment includes land proposed for the Calvert to Kagaru segment of the Australian Rail Track Corporation's Melbourne to Brisbane Inland Rail project. Secondly the forecast population growth of the broader city of Ipswich from the current approximation of 233,302 to 435,000 by the year 2031.[86] Damming land suitable for housing part of those forecast extra 201,698 persons will necessarily create public policy and planning problems elsewhere within the Ipswich City Council's local government area which are not conveniently serviced, as the areas traversed by Warrill Creek are, by a major road – in this case the Cunningham Highway linking Ipswich and Warwick.

Newman, in his post-premiership capacity recounted earlier in this chapter as a Liberal Democrats Senate candidate for the 2022 federal election, cited his government's December 2014 feasibility study, foreshadowing that '*if elected as a Queensland senator he would campaign to have Queensland and Australian governments re-evaluate new dams in the Brisbane Valley*'.[87]

The '*Brisbane Valley* ' reference suggests the second of the two proposals identified in the study handed to Newman's then government during December 2014: to dam the Upper Brisbane River at Linville, upstream, by quite some distance, of Wivenhoe Dam. However after her unlikely victory not even two months after the study was commissioned, when Annastacia Palaszczuk wrested the premiership from Newman on 31 January 2015, Palaszczuk ordered no further investigations into the Linville proposal, instead focus-

---

85 Ibid.

86 Ipswich City Council commissioned '*Flood Review 2022 – Strategic Review Report and Operational Review Report*' tabled 29 November 2022 ('Ipswich City Council 2022 flood review'), p. 7.

87 Tony Moore, 'Dam or be damned: Campbell Newman revisits 2014 plan for flood protection', *The Brisbane Times*, 12 May 2022.

sing on upgrades to the extant Somerset, North Pine and Wivenhoe Dams with a timeframe for completion until 2035.[88]

Irrespective of whatever reasons, or non-reasons, existed for the dispensing of the Linville proposal, it is independently argued that no dam at Linville would have any meaningful flood mitigation properties. Linville is located not only too far north in the Upper Brisbane River catchment, it is located too far north-west. The significance of the westerly aspect is the inherent propensity of rainfalls to dissipate the further west one travels in the overall Brisbane River catchment:

> *There is a general east-west rainfall gradient across the catchment primarily due to the effects of coastality and orographic uplift ... precipitation at Cleveland exceeds that at Ipswich (about 50km west) by 30 to 50 percent during the rainy months; the Stanley River (a northeastern tributary* [of the Brisbane River]*) has a runoff per unit area four times higher than Lockyer Creek (a western tributary).*[89]

One genuine example of the extremity of rainfalls over the Stanley River/Somerset Dam areas in the demonstrable north-east – as opposed to north-west – of the Brisbane River catchments is the quantum of the rainfall which caused the 1974 floods:

> *Wanda, a capricious weather system ... dashed across the coast near Double Island Point north of Brisbane. Wanda quickly degenerated into a rain depression which sat over the Stanley and Brisbane River catchments. More than 642 millimetres of rain fell on Brisbane in just over thirty-six hours. Somerset Dam held some of the water in check, but had little influence on the torrent which poured down the Brisbane and Bremer Rivers.*[90]

---

[88] Lydia Lynch, 'Deluge defence report shelved', *The Australian*, 10 March 2022, p. 5.

[89] *The Brisbane River: A source-book for the future*, Eds Peter Davie, Errol Stock and Darryl Low Choy, The Australian Littoral Society, 1990, p. 3.

[90] Helen Gregory, *The Brisbane River Story – Meanders through Time*, Australian Marine Conservation Society Inc, 1996, p. 121.

Wanda's impact on the Stanley and Brisbane River catchments being illustrated by this track map[91]:

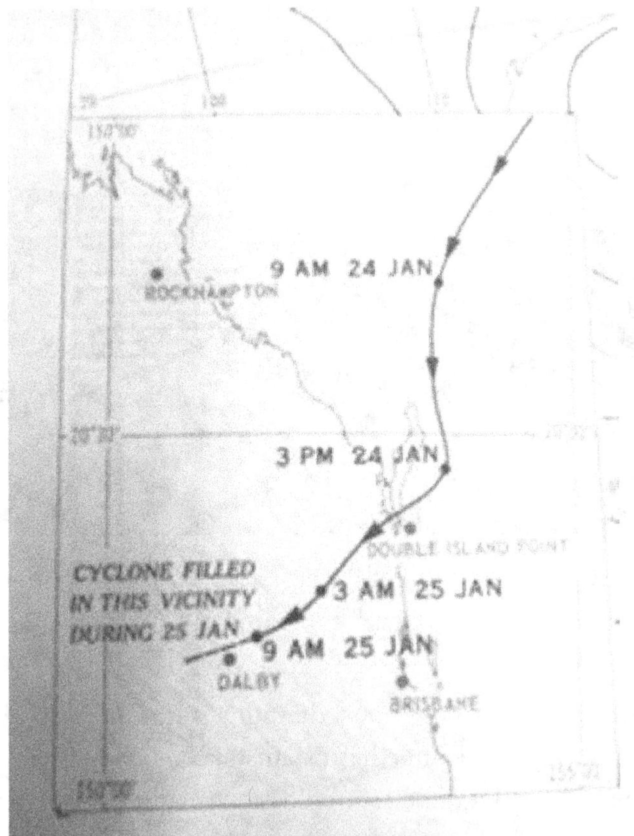

The falls which contributed to the 1893 floods also began their long march to Brisbane in the Stanley River catchment:

> [Somerset] *dam was named for an early grazier in the area, Henry Plantagenet Somerset, who had desperately, but unsuccessfully, attempted to warn Brisbane of the approach of the 1893 flood.*[92]

Linville, on the other hand, is north-west of the Stanley River and its associated Somerset Dam, both of which possess a consequentially greater propensity to receive substantial rainfalls.

---

[91] Department of Science Bureau of Meteorology, *Report by Director of Meteorology*, Australian Government Publishing Service, Canberra, 1974, p. 19.

[92] Helen Gregory, op. cit., p. 81.

Secondly, the Brisbane River, being as close to its source as it is at Linville, is more like a stream:[93]

Thirdly, and as a result of the second point, Linville's total catchment area percentage is 15%,[94] compared to Wivenhoe's vastly greater 51.584%.

In short Linville would be a dam much less likely to fill and, in any event, given its far-northern status in the Upper Brisbane River catchment, even if it did fill it would dam too little water to have any flood-mitigation impact of any benefit.

The Queensland Government map on the page opposite, which also displays a potential location for a dam based on Warrill Creek, vividly illustrates this.[95]

Finally there would be the inherently fraught political context which would bedevil any dam proposal within South-East Queensland due to population growth. Queensland's population is substantially greater than it was during the 1968-1987 Sir Joh Bjelke-Petersen era when his many constructed dams would have involved

93 Mattinbgn. Wikimedia Commons Attribution-Share Alike 3.0 Unported (CC BY-SA 3.0).

94 *Brisbane River Flood Study*, Volume 1, Final Report: Brisbane City Council, 1999, p. A4 appendix 3.

95 'Linville Dam Proposal Shortlisted', Southburnett.com.au, 16 December 2014.

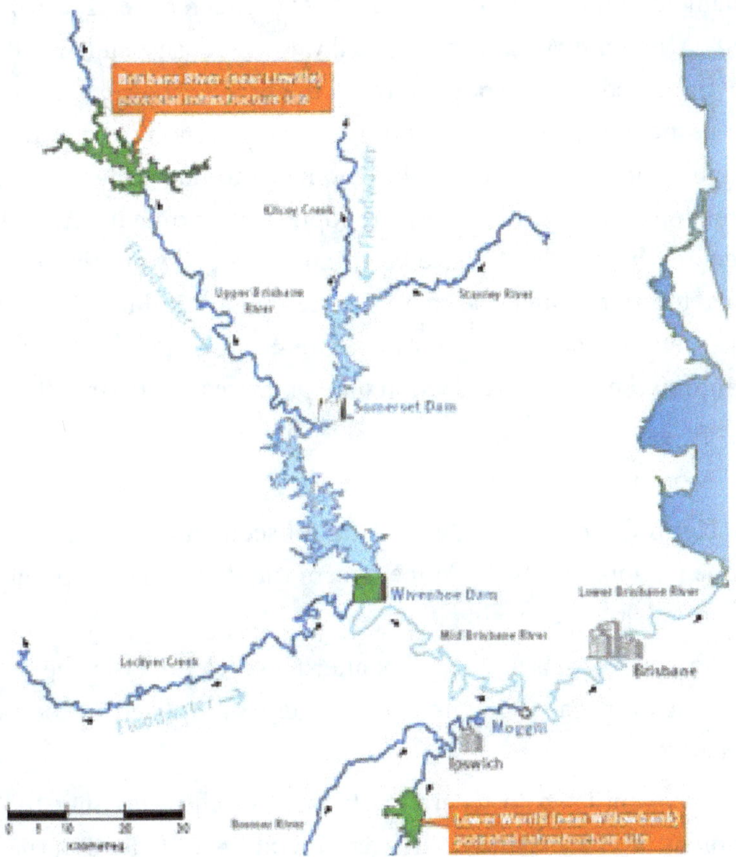

minimal rendering homeless of residents. Or, to the extent that people were forced by resumptions to have had to have vacated, the cult of personality that followed Premier Bjelke-Petersen, keeping him in power for 19 years having never lost a single election as premier, sufficiently immunised him from any political backlash.

That midas touch however could not be relied upon by a subsequent premier named Russell Cooper. Though the previously cited Wolffdene proposal was shelved in the immediate aftermath of the 1974 floods, over time it re-emerged and by 1989 then Queensland Premier Russell Cooper determined to reinstate and complete the proposal. Not that the population to be impacted were as sanguine as the 350 property owners required by Premier Bjelke-Petersen's government to permanently depart for Wivenhoe's creation were; what changed in the just under two decades between the early 1970's and 1989 when the Wolffdene proposal was raised, suspended and then re-enlivened was the population of the broader Logan City council area, including Wolffdene and its many surrounding suburbs. The population rose as housing developments to home the increasingly growing South-East Queensland population sprung up. All of those persons' homes would have been inundated; all would have had to have found housing elsewhere. Hence the political backlash, which the then opposition leader Wayne Goss utilised to his advantage by winning the 1989 state election on a promise to permanently shelve the project.

A promise Goss kept.

Nor did Premiers of Queensland subsequent to Goss, even of the same National Party as Cooper, deem the Wolffdene proposal worthy of raising ever again either.

So why then, a full 18 years after Goss' victory, did the by now ancient proposal suddenly come to be re-ventilated in the public debate?

And ventilated in federal parliament of all places, not the parliament of Queensland. Why were members of federal parliament from NSW and Victoria[96] suddenly raising an issue local to South-East Queensland in the nation's parliament?

As addressed in *Tennyson Breach* in 2012, much print room in federal parliamentary Hansard was devoted, during 2007, to a desperate bid by the incumbent Howard Government to discredit then ALP Opposition Leader Kevin Rudd in the hope, futile in the result, of preventing his ascension to the prime ministership.

Yes but Rudd didn't cause South-East Queensland to descend into such a prolonged drought? No one person could. What was the back-story?

Rudd's role in 1989 as a senior public servant in the incoming Wayne Goss state government of Queensland was dredged back up

[96] *Tennyson Breach*, pp. 73 & 74.

due to Goss' promise in the 1989 Queensland state election campaign not to build the Wolffdene Dam.

Grounding epithets that Rudd was '*the de facto premier of Queensland*' and '*the architect of south-east Queensland's drought*'.[97]

The latter of which was a quote of then Minister for the Environment and Water Resources Malcolm Turnbull. Turnbull, eight years later, also, like Rudd, ascended to the nation's highest office, yet did not cause one single dam to have been commissioned during a three year prime-ministership. '[H]*e* [Turnbull] *was a compelling-but-failed national leader*'[98] opined journalist Aaron Patrick, and though Turnbull's sudden silence on the issue of dams he was so voluble about in 2007 was not directly cited by way of example in Patrick's work, the juxtaposition remains glaring.

Proving the premise of this chapter: that while dam proposals are inherently governed by and can only meaningfully be evaluated according to precepts of engineering and science, the politics of dams will always be noisier and garner more attention, despite, or perhaps because of, the abject lack of any consensus.

Save, dare it be asserted, that the forecasts made by Dr Cook in 2019 and myself in 2012 have already come true so far as the concept of permanently flood-proofing Brisbane is concerned, and will so prove in the future again.

---

[97] Ibid., p. 74.

[98] Aaron Patrick, *Ego Malcolm Turnbull and the Liberal Party's Civil War*, Harper Collins, 2022, p. 301.

5

The 2011 Floods Class Action

Sub Ch 5.1 – The litigation and the politics of the litigation

In today's litigious world it was not altogether unsurprising that a class action lawsuit was commenced following the 2011 flood.

Emotions were raw in the immediate post-2011 aftermath; many were quick to lay blame at the operators of Wivenhoe Dam, contending that had it not been mis-managed it would have 'done its job' and prevented the Brisbane River flood of 2011.

Perhaps due to the accepted vulnerability of the then Wivenhoe-less Brisbane in 1974, no formal commission of inquiry helmed by a judicial figure took place. Rather what occurred was an '*engineering led review ... a quickly convened symposium some three months after the event in April 1974*'. The differences between the post-event procedures in 1974 and 2011 were reported by commentators who recalled both flood events this way as '*notable*':

> *One notable difference between the 1974 flood symposium and the Queensland Floods Commission of Inquiry was that in 1974 there was not the legal interest regarding provision in legislation for protection against liability of dam operations.*[99]

Of course, 1974 was a generation ago compared to 2011. Community attitudes which formerly eschewed recourse to litigation had tilted quite the opposite way 37 years later. Also of difference was the rapid growth in the 'no win, no fee' access to justice for plaintiffs legal business model.

So it therefore came to pass that a lawsuit emerged: in 2014 a

---

[99] Rob Ayre, Terry Malone and John Ruffini, 'Wivenhoe, January 2011: The Dam Truth', The Munro Oration, presented at Engineers Australia's 2022 Hydrology and Water Resources Symposium ('Ayre, Malone & Ruffini'), p. 25.

small business which had suffered loss as a result of the 2011 flooding, Rodriguez & Sons Pty Ltd ('Rodriguez'), commenced a class action seeking to recover damages from the State of Queensland, and two statutory agencies then responsible for operating Wivenhoe Dam: Queensland Bulk Water Supply Authority trading as Seqwater ("Seqwater") and SunWater Ltd[100] ("SunWater") – collectively "the operators".

Rodriguez was a sports store operator in Fairfield Gardens, a shopping centre in Brisbane's inner southern suburb of Fairfield. As excess waters from the swollen Brisbane River flowed in an easterly direction away from the River and into the suburb of Fairfield, homes, businesses and Fairfield Gardens were inundated. Including Rodriguez' sports store.

Incidentally Fairfield Gardens was identically affected by the 2022 flood, closing on the afternoon of Saturday 26 February[101] and not being able to re-open for one week thereafter. That centre contained the ward office for the local Tennyson Ward Brisbane City Councillor Nicole Johnston during both 2011 and 2022. Councillor Johnston asserted that '*there was no functioning ward office for the second time* [in 11 years] *in a major natural disaster. This is unacceptable*'.[102]

Proving that while the 2022 event was markedly different from 2011, there were many almost precise similarities as well, especially adjacent to or otherwise influenced by the Brisbane River.

Rodriguez' allegation was that, in the early days of 2011, the operators negligently failed to commence releasing water from Wivenhoe Dam several days earlier than they did, in anticipation of heavy rainfall, so as to avoid the need to release much larger volumes in

---

[100] *Queensland Bulk Water Supply Authority t/as Seqwater v Rodriguez & Sons Pty Ltd* [2021] NSWCA 206 (8 September 2021), at [4] – from here on in the footnotes this decision of the NSW Court of Appeal will be cited simply as '*Appeal*'.

[101] Fairfield Gardens' official facebook page.

[102] Nicole Johnston, Submission to the de Jersey 2022 Brisbane Flood Review, 8 April 2022, p. 13.

a short period after the rainfalls which caused large inflows and a surge in Wivenhoe Dam's level.[103]

The trial judgment of 29 November 2019 {defined in this work as 'First instance'} of the trial judge – Justice Beech-Jones who was referenced in Chapter 2 – was the direct result of the trial of Rodriguez' class action. A trial that ran for 130 days. The first-instance judgment, as opposed to the subsequent appeal, was a victory for Rodriguez and his many co-joined claimants, all of whom suffered varying degrees of loss and damage as a result of the 2011 flooding. The operators lost; Seqwater was adjudged to be 50% liable to the cohort of claimants; the State of Queensland and SunWater shared the remaining 50% of the liability, 30% to SunWater and 20% to the State.[104]

Given the sheer quantum of the potential damages bill, and the unique nature of the case, it was not altogether surprising that the operators would have wished to have appealed the trial verdict.

Seqwater in fact did.

However the State of Queensland, both in its own right and in terms of the company it owned all of the shares in, SunWater, chose not to appeal. What was interesting about this decision was that the State of Queensland also owned all of the shares in Seqwater, with the same shareholding Ministers, a state of affairs that still exists in 2022.[105]

Of course the State of Queensland had another consideration other than money that most litigants do not have: the verdict of the other court, the court of public opinion. Hence the decision not to appeal. Perhaps in hindsight not the most prudent of legal decisions, though with the primary judgment handed down within 1 year of a constitutionally mandated Queensland state election date of 31 October 2020, a compassionate approach from an incumbent govern-

---

[103] Appeal, at [2].

[104] Appeal, at the fifth un-numbered paragraph of the headnote.

[105] Ayre, Malone & Ruffini, p. 3.

ment seeking re-election definitely had merit when viewed through a prism of retail politics.

Even if, ironically, no-one in those final months of 2019 could have foreseen what subsequent public policy challenge was then to emerge barely after the new years eve parties to ring in the new year had wrapped up: the subsequent dislocation and disruption caused by the truly unprecedented Corona Virus pandemic.

The pandemic was the story of 2020. Flood class action – what flood class action? The dire warnings of the impact of the virus and subsequent lock-downs confining persons to homes with exits allowed purely for limited reasons such as socially-distanced exercise and shopping for essential items only meant that by 31 October 2020 none but the lawyers involved and the claimants themselves would have even remembered the post-trial result of the floods class action, let alone allowed it to influence their voting thinking.

Therefore had the State of Queensland and SunWater taken the hit politically in the 28 day appeal period after the trial decision was handed down and chosen to have joined Seqwater in having appealed, much in the way of damages would have been saved with no net negative result election wise. The 2020 election will of course be remembered for the ways in which 'Palaszczuk's Pensioners', some for the first time ever in their lives, combined to vote Labor and return the incumbent Government with an even greater majority as a thank you vote for 'keeping us safe'.

The decision not to join Seqwater's appeal thereby providing the worst of both worlds: a large spend with no public kudos.

Admittedly the State of Queensland and SunWater could not have known the result of any appeal. No litigant ever does.

Nor, of course, could the incumbent government have known that the inevitable political backlash from appealing would have exhausted away entirely by the pandemic which became the only issue of any substance during the actual year of the election.

Sub Chapter 5.2 – The Brisbane River as an overall system

To begin to comprehend the myriad of facts and principles of law that were decided and determined in both stages of Rodriguez' class action, consideration needs to be taken of exactly what the Brisbane River is and entails, including of course Wivenhoe Dam and where it sits relative to the overall system.

The total Brisbane River catchment area is vast; its total catchment has an area of some 13,570km2[106] stretching from a latitude congruent with Maroochydore in the north to almost as far as the Queensland and New South Wales border in the south. In turn there are five main sub-catchments: (a) Upper Brisbane River (above Wivenhoe Dam); (b) Stanley River (above Somerset Dam, which in turn is above Wivenhoe Dam); (c) Lockyer Creek; (d) Bremer River (including Warrill and Purga Creeks); and (e) the Lower Brisbane River[107]:

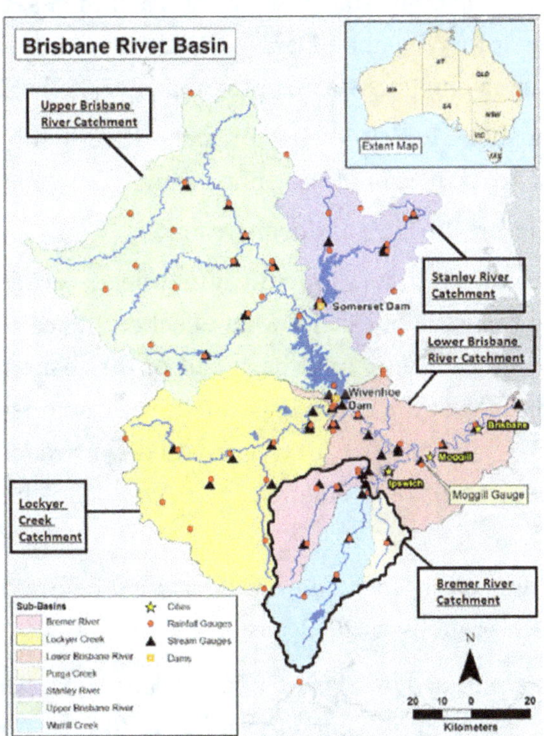

[106] *First instance*, Chapter 2, at [39].

[107] Ibid.

In turn the Lower Brisbane River sub-catchment contains a sub-sub-catchment not mentioned in this image: the Oxley Creek system. Oxley Creek feeds into the Brisbane River at the suburb boundary of Graceville and Tennyson, and forms those suburbs' boundary. A creek it may be named but the reality is that Oxley Creek is a significant system with a 522km long stream network and a total area of 258 square kilometres. The many puddles of mud depicted so strikingly on the cover of my previously cited tome *Tennyson Breach* were from that very sub-sub-catchment. And also, though not visible yet troublesome nonetheless, was the way that Oxley Creek during January 2011 was '*slick with contaminants and littered with industrial debris. Much of this pollution was discharged from the many industrial and commercial facilities located within the Oxley Creek catchment*'.[108]

Of the five major Brisbane River sub-catchments, the two that directly flow into Wivenhoe Dam – the Upper Brisbane River and Stanley River catchments – are approximately 7,000km2 in area.[109]

Meaning that 51.584% of the Brisbane River's sub-catchments include and are upstream of Wivenhoe Dam.

And therefore 48.416% is downstream.

Even at this preliminary juncture it may be discerned that, when considering the issue of exactly what effect existence or non-existence of Wivenhoe Dam could have had on Brisbane in 2011, the fact that 48.416% of the entire Brisbane River catchment up to and including Brisbane CBD's City Gauge was downstream, and hence wholly independent of Wivenhoe Dam, could not have been ignored.

As pronounced in *Tennyson Breach* as long before the first shot fired in the class action as 2012 was:

> [T]*he devastation of the January 2011 floods had a demonstrable eastward trend: first the sheer terror of lives and properties*

[108] QFCI Final Report, p. 153, footnotes omitted.

[109] *First instance*, Ch 2, at [40].

lost in Toowoomba, Grantham and the Lockyer Valley west of Brisbane on Monday 10 January 2011, followed by the severe flooding that impacted Ipswich and Goodna to those areas' east on Tuesday 11 and Wednesday 12 January 2011, and, as those waters inexorably flowed further east to the city of Brisbane, Brisbane itself by the peak day of Thursday 13 January 2011.

A very cursory view of a map of South-East Queensland will soon make clear that Wivenhoe Dam had zero impact at all in holding back those floodwaters for the simple reason that the waters added to the Brisbane River from Toowoomba and the Lockyer Valley were conveyed via the Lockyer Creek whilst the flood waters from Ipswich were conveyed via the Bremer River, both of which intersect with the Brisbane River downstream from Wivenhoe.[110]

A similar conclusion was found by, and instrumental in the reasoning of the NSW Court of Appeal, which will be examined in sub-chapter 5.4 below.

Finally, and perhaps most visionary given that this statement was published 26 years ago, Helen Gregory described causative factors of the 1974 floods this way: '*Wanda quickly degenerated into a rain depression which sat over the Stanley and Brisbane River catchments ... Somerset Dam held some of the water in check, but had little influence on the torrent which poured down the Brisbane and Bremer Rivers*'.[111] The Brisbane River, of course, being wholly undammed in 1974; Somerset Dam dammed only the Stanley River.

Sub-chapter 5.3 – The Dual Roles of Somerset and Wivenhoe Dams

Chapter 4 essayed the gestational history of Wivenhoe Dam including a demonstrable pretext of drinking water supply for a rapidly growing city of Brisbane that Somerset Dam alone could not guarantee into the future.

---

110 *Tennyson Breach*, p. 116.

111 Helen Gregory, *The Brisbane River Story – Meanders through Time*, Australian Marine Conservation Society Inc, 1996, p. 121.

The NSW Court of Appeal performed a similar exercise:

Although the litigation emphasised the dams' flood mitigation purpose, the dams [Somerset and Wivenhoe] *also served the purpose of supplying water to Brisbane and south-east Queensland. This may be seen in the title of the report presented to the Queensland Parliament in 1934 by the Special Committee which was "appointed to Investigate and Report upon Brisbane Water Supply and Flood Prevention".*[112]

The Court then noted the obvious potential for conflict:

Those two principal purposes are opposed to each other. Each dam has a finite capacity to hold water ... That capacity can be used to store water for consumption. Alternatively, the dam may be kept empty, available for the temporary storage of flood waters. Maximising water storage would lead the dam to be kept as full as possible for as much time as possible. Maximising the dam's capacity for flood mitigation would cause the dam to be left empty for as much time as possible.[113]

The Court's use of the term '*finite*' is more than appropriate when considering the theme of Chapter 1 of this book: that though the rainfalls which fell in the Brisbane River's two uppermost sub-catchments during 2011 and 2022 were extreme, just as the 1893 and 1974 rainfalls also were, those events were definite outliers when considering how little rain falls in those catchments during the phases below the line depicted in the '*Qld rainfall – past, present & future*' graph.[114] Including the commensurate decrease in overall terms of Wivenhoe Dam's capacity from 188.7% at the height of the 2011 flood event to a very worryingly low 36.3% over the course of a period of time as short as the remainder of that decade. The alarming nature of which having been depicted so graphically by the Bureau of Meteorology.[115]

[112] Appeal, at [162].

[113] Appeal, at [163].

[114] Supra, fn 3.

[115] Supra, fn 8.

Hence the NSW Court of Appeal's concept of the need for '*some mechanism for regulating the compromise between the dam's water storage purpose and its flood mitigation purpose*'.[116] That mechanism is the Flood Operations Manual, the content of which was examined at great length in both the trial and the subsequent appeal, and will be in sub-chapter 5.4 below.

In the meantime, the duality of Wivenhoe Dam needs to be further explained. Wivenhoe's maximum capacity is 3.132 million megalitres.[117] Sydney Harbour's is 500 gigalitres,[118] or 500,000 megalitres, making Wivenhoe's full capacity well north of six Sydney Harbours.

On any objective basis, at maximum capacity it is one vast reservoir of water.

In turn, this concept of 'full capacity' is in fact the sum total of two separate capacities, both of which reflect the dualities referred to by the NSW Court of Appeal. The capacity of the Dam's urban water supply compartment – that is water supply for Greater Brisbane and surrounds – is 1,165,000ML,[119] or about 37% of the total 3.132 million megalitre overall storage. In terms of physical representation at Wivenhoe's dam wall face, the upper limit of the flood compartment's capacity – defined as 'FSL' {full supply level} – is reflected in a wall height of 67.0m Australian Height Datum (AHD). Wivenhoe Dam's remaining volume exceeding the FSL, that is between 67.0m AHD and the maximum water level height which must never be exceeded of 80.0m AHD lest the Dam overtop,[120] comprises the flood mitigation compartment. With a total capacity of 1,966,000ML,[121]

---

[116] Appeal, at [164].

[117] Michael Madigan, 'Flood fears build as 'lessons not learnt'', *The Courier-Mail*. 2 September 2019, p. 15.

[118] John Donegan, '11 Things you should know about Sydney Harbour', *ABC Radio* Sydney, 4 September 2014.

[119] Dr Luke Toombes and Rob Ayre, 'Resetting Expectations of Brisbane's Flood Protection', ISBN 978-1-925627-64-0, presented at Engineers Australia's Hydrology and Water Resources Symposium 2022, p. 3.

[120] The implications of an overtopping being discussed in sub-chapter 5.4 below.

[121] Supra, fn 117.

the flood mitigation compartment is therefore greater, by quite some way, than the 1,165,000ML full supply level. Proving that, in practice, the opposing two principal purposes are not reflected in identical water storage totals. If anything, a degree of prioritisation can be discerned to be given to the flood mitigation imperative over that of drinking water storage.

Therefore there will be repeated references in this book to these dual compartments of Wivenhoe Dam, including an imperative to ensure that the 'gap' between the full water supply compartment level of 67.0m AHD is never utilised during phases of dry weather.

Sub Ch 5.4 – '*The Manual*'.

Notwithstanding the pre 1974 flood era, when water supply was the primary focus of the then planned dam that subsequently became Wivenhoe, its focus radically changed when the 1974 flood occurred. For that reason, Section 32 of the *Wivenhoe Dam and Hydro-Electric Works Act 1979* (Queensland) ('Wivenhoe Act') mandated the preparation of '*a manual of operational procedures in relation to each reservoir ... for the purpose of flood mitigation pending completion of the Wivenhoe dam project*'.[122] The *Brisbane and Area Water Board Act 1979* (Queensland) required the preparation of a manual of operational procedures, and expressly preserved the operation of the manual prepared under the Wivenhoe Act, for the post-construction phase when Wivenhoe Dam opened in 1985, until that manual had ceased to be effective.[123]

Hence the phrase that became ingrained into the lexicon after the 2011 flood event, when the '*Manual of Operational Procedures for Flood Mitigation at Wivenhoe Dam and Somerset Dam*' ('the Manual') was exhaustively analysed:

- Firstly by the Holmes Commission of Inquiry established by the Bligh Government in the immediate months after

[122] Appeal, at [166-167].

[123] Appeal, at [168].

January 2011, which in turn was extensively reported in the media, as commissions of inquiry tend to be;

- Secondly by the NSW Supreme Court in the trial decision;
- Thirdly by the NSW Court of Appeal; and fourthly and finally by
- The High Court of Australia on 12 April 2022, when each party had a minimum of 20 minutes only to attempt to convince the High Court to agree to hear an appeal from the NSW Court of Appeal judgment – see sub chapter 5.10 below.

During Wivenhoe Dam's existence various revisions to the Manual occurred; the final revision of the Manual before January 2011 was approved on 22 December 2009, which version was therefore extant as of the 2011 flood event.[124] That version provided five objectives in order of importance, namely:

- Ensure the structural safety of the Somerset and Wivenhoe dams;
- Provide optimum protection of urbanised areas from inundation;
- Minimise disruption to rural life in the valleys of the Brisbane and Stanley Rivers;
- Retain the storage at Full Supply Level at the conclusion of the Flood Event; and
- Minimise impacts to riparian flora and fauna during the drain down phase of the Flood Event.[125]

In addition, Section 3 of the Manual, titled '*Flood Mitigation Objectives*', identified these five objectives explicitly listed in the same descending order. The catastrophic consequences of the structural failure of either the Somerset or Wivenhoe Dams were emphasised in section 3.2 of the Manual: the structural safety of both dams

124 Appeal, at [171].

125 Appeal, at [182].

'*must be the first consideration*' in flood mitigation operations.[126] This follows because '[T]*he dams were built in such a way that if waters rise close to or exceeding the dam crest, there is a chance the structure will fail. That would cause catastrophic flooding in Brisbane and Ipswich*.'[127]

That's putting it mildly when one considers the statement made in the immediately preceding sub-chapter that Wivenhoe Dam's maximum capacity is 3.132 million megalitres; more than six Sydney Harbours. This therefore puts beyond doubt that, as a result of Wivenhoe Dam's status as a zoned earth and rock fill embankment dam, it can not, and must not, overtop; if it does the entire structure collapses.

Another imperative against overtopping is the fact that on the top of the dam wall resides an actual highway. Highway 17, the Brisbane Valley Highway, shares the dam wall and comprises its ceiling. This is that Highway's westbound approach {away from Brisbane} to the commencement of Wivenhoe's dam wall at Cormorant Bay:

While this picture depicts a mere part of the vast dam wall, this time driving from the western side back towards Brisbane; the max-

126 Appeal, at [187] and [189].

127 Appeal, at [160].

imum height which must never be exceeded of 80.0m AHD is represented by the top of the concrete wall on the left road shoulder; the road deck itself is approximately 79.1m AHD[128]:

A dam wall that both holds back water and permits vehicles to traverse from either side. Versatile!

One thing that is anything but a joking matter though is what would happen if those at least six Sydney Harbours worth of water capable of being held back behind Wivenhoe's lengthy dam wall were able to sweep upwards and then overtop. Studies by engineering firm URS in 2013 and 2014 concluded that the failure of Wivenhoe Dam would put at risk almost 300,000 people, kill about 400, and occasion a total cost to the community of up to $100 billion.[129]

Hardly surprising then that the Manual prioritised, in excess of all other considerations, the preservation of Wivenhoe Dam's structural integrity. Summarised by the Appeal Court in this descending order:

1. First, and above all else, each dam must be managed to prevent a catastrophic structural failure.
2. Secondly, so long as there is no threat to the structural integrity of the dam, the main flood mitigation purpose is to prevent urban inundation downstream.

---

[128] Rob Ayre, interview with the writer.

[129] Mark Solomons, 'Plan to increase Wivenhoe by two Sydney Harbours in $900m safety revamp', *The Brisbane Times*, 4 April 2017.

3. Thirdly, there must be some mechanism for regulating the compromise between the dam's water storage purpose and its flood mitigation purpose.[130]

As a result, the metrics assume crucial importance:

> [T]*he spillway of Wivenhoe Dam had a crest at an elevation of 57m. Above the crest were five radial gates which could be raised to allow the release of water over the spillway. Full Supply Level for Wivenhoe was 67m. The top of the core of the dam was 80m but was described as "not resistant to overtopping".*[131]

The concept in this quote of '*Full Supply Level*' is the full supply of Wivenhoe Dam's drinking water storage compartment concept mentioned in sub-chapter 5.3.

Thus, once the dam level reached 80m there was an expectation of a structural failure with potentially '*catastrophic consequences*'.[132]

The only objection this writer has is to the word '*potentially*'. If Wivenhoe Dam did collapse, the catastrophic consequences would be certain, not a mere possibility.

Indeed, the Appeal Court dropped the epithet '*potentially*' and replaced it with a much more certain '*plainly*' some 202 paragraphs later in its judgment: '*Plainly it would be catastrophic were the dam to fail*'.[133]

This therefore explains the discerned prioritisation in sub-chapter 5.3 by virtue of the flood mitigation compartment being greater in volume than that for water storage:

> *Flood magnitudes cannot be predicted with any great reliability, and a strategy based on regularly using the full volume would have extreme risk.*[134]

---

[130] Appeal, at [164].

[131] Appeal, at [29], footnotes omitted.

[132] Appeal, at [29], footnotes omitted.

[133] Appeal, at [231].

[134] Dr Luke Toombes and Rob Ayre, *Resetting Expectations of Brisbane's Flood Protection*, p. 7.

Not that the idea of intruding into the flood mitigation compartment by permanently increasing the water storage compartment to a volume exceeding 1,165,000ML [or 67.0m AHD] has not been ventured though. On 9 March 2010, before the 2011 flood event and in a context of memories of the Millennium Drought being both very recent and very raw, then Deputy Opposition Leader of Queensland, Jeff Seeney, advocated in parliament for exactly that to happen:

> *... it is time to review the comparative uses of the total storage available in Wivenhoe Dam ... more of the total available storage space should be used for water storage so that there is less chance of a water crisis developing in the future. Obviously the dam needs to continue to be operated to maintain a flood buffer for Brisbane and any decision about storing extra water will have an impact on the effectiveness of that dam's flood buffer role, but this is a question of finding the right balance between the two roles that Wivenhoe Dam perform ... I call on the minister today to ensure that no water is released from Wivenhoe Dam and that serious commitments are made to developing this proposal to increase the storage level ... The water that is running into the dam at the moment may well be sorely needed in the future to avert another water crisis.*[135]

Not all together surprisingly, Seeney's idea has not been revisited; to the contrary, and as will be examined in Chapter 6 below, following both the 2011 and 2022 flood events, exhortations to politicians have urged the polar opposite – '*massive*' pre-emptive releases from the water storage compartment, effectively reducing, quite savagely, its 1,165,000ML upper limit.

With reference back to the '*catastrophic*' nature of an overtopping event, and in an attempt to provide some reassurance to any of the almost 300,000 members of the public estimated by URS as being at risk reading this book quite apprehensively at this stage, an

[135] Hansard, Qld Parliament, 'Matters of Public Interest', 9 March 2010, p. 650.

insurance policy of sorts is built into Wivenhoe Dam to automatically prevent overtopping. Namely, fuse plug spillways. Below the critical '*not resistant to overtopping*' level of 80m exist '*three "fuse plug" embankments designed to erode once water flowed over them. Although the erosion was intended to occur in a "controlled manner", erosion would result in a large and uncontrolled discharge of water. The "trigger levels" for the erosion of the three fuse plug embankments were stated as 75.7m, 76.23m and 76.78m respectively*'.[136]

This self-taken picture from the Lake Wivenhoe Information Centre parklands looking south depicts those very fuse plugs, constructed in 2005:

The semi-trailer depicted is traversing the very same Brisbane Valley Highway depicted in the two previous dam wall photographs. The difference is that the Highway at that point bridges that part of the dam as a result of the dam wall being excised and replaced with a bridge so as to dislocate the Highway from the overall dam structure in the event of the fuse plugs eroding as designed.

Half a minute later, that same semi-trailer had advanced further eastwards along the Brisbane Valley Highway, as depicted here:

---

136 Appeal, at [30].

The reason for this particular picture is to visually demonstrate not only the sheer length of the Dam wall, but also what the flood mitigation compartment actually looks like at the Wall face; as of June 2022, Wivenhoe Dam's drinking water storage compartment was at 90% capacity, meaning that the daylight visible between the height of the water and the permanent concrete guard rail on the Highway's left shoulder portrays the sum total of 10% of the water storage compartment plus all of the flood mitigation compartment.

The political arguments therefore being the Seeney view, which would have the height of the Dam equating to the upper limit of the drinking water compartment marching higher up the wall, while the advocates of '*massive*' pre-emptive releases would have the water level much lower so as to provide for greater capacity to absorb inflows.

Putting the politics to one side in lieu of the more prosaic, the Manual fixes 67m AHD as the upper limit of the water storage compartment. There is therefore a balancing act required between the 67m full supply level of that compartment and the bottom of the three ranges where the first of the three Wivenhoe fuse plugs would, theoretically, erode and trigger an automatic release:

> *The area above 67m was identified in the Manual as "temporary flood storage". However, to avoid uncontrolled discharge of water* [by triggering of the first fuse plug], *it was necessary to keep the water level below 75.7m.*[137]

The upshot being that, at levels as high as 75.7m, 76.23m and 76.78m respectively, if the fuse plugs work as they are intended, the good news is that Wivenhoe Dam will not collapse.

The bad news is that when the fuse plugs erode away after Wivenhoe Dam water heights exceed 75.7, 76.2 and 76.7m, the consequence is a release of '*substantial flows of water in an uncontrolled way down three separate spillways until the depth returns to 67m*'.[138]

The image on the opposite page is of Kwa-Zulu Natal's Shongweni Dam which did, during April 2022, reach the requisite level to cause its own fuse plugs to trigger.[139]

As stated in chapter 2, 341 died during the Durban flood event of April 2022. The ferocity of those waters uncontrollably being released from Shongweni Dam making the loss of life, tragic as it of course is, explicable especially to the extent of a want of defences for persons and residents downstream.

---

[137] Appeal, at [31].

[138] Appeal, at [197].

[139] Grant Shimwell, University of Natal BSc Engineering (Civil), Civil Engineering BSc 1964-1968, via LinkedIn.

The '*substantial uncontrolled*' flows which a fuse plug trigger would cause to Wivenhoe Dam were explained this way:

> *Thus if the ... first spillway plug is eroded, then the entire flood capacity of the dam – the volume of water between 67m and 75.5m (1,160,000Ml) – will be released down the first spillway, starting at a rate of around 1873m3/s, and thereafter reducing as the water level is lowered. If the water level reaches 76.5m and the second spillway plug is eroded as well, then the same will occur, save that this time the water will be released ... in excess of 5400m3/s – which will of itself ... cause damaging floods to urban areas of Ipswich and Brisbane.*[140]

This is because the critical threshold cubic metreage per second of water, understood by the Manual extant as of January 2011, was 4000 at Moggill: Brisbane River's second-most southerly point before it begins its north-easterly turn towards the centre of Brisbane and ultimately its exit to Moreton Bay. The Manual specified that a Brisbane River flow rate of 4,000 cubic metres of water per second at Moggill was the upper limit of flows that would not damage homes and properties.[141]

---

140 Appeal, at [198].

141 Appeal, at [627].

In terms then of practical application of the fuse plug triggering water levels:

> *The water levels of 74m and 68.5m referred to predictions used to identify the primary objective and the particular strategy to be deployed. If predicted dam levels were to exceed 74m, then the primary objective was to preserve the dam. If predicted dam levels exceeded 68.5m, then the primary objective was to prevent urban inundation. If the predicted dam levels were lower, then subordinate flood mitigation objectives (such as preserving the downstream bridges) came to the forefront.*[142]

The reference to '*objectives*' heralds the Manual's Section 8.4, headed '*Flood Operations Strategies*', which in turn identified four '*strategies*' for Wivenhoe known as W1, W2, W3, and W4:

> Strategy W1 – the Primary Consideration is Minimising Disruption to Downstream Rural Life.
>
> Strategy W2 is a Transition Strategy where the primary consideration changes from Minimising Impact to Downstream rural life to Protecting Urban Areas from Inundation.
>
> Strategy W3 – The primary consideration is Protecting Urban Areas from Inundation.
>
> Strategy W4 – The primary consideration is Protecting the Structural Safety of the Dam.[143]

In turn the Manual provided that the '*intent*' of each of these four strategies was:

1. The intent of Strategy W1 is to not to submerge the bridges downstream of the dam prematurely.
2. The intent of Strategy W2 is limit the flow in the Brisbane River to less than the naturally occurring peaks at Lowood

142 Appeal, at [227].

143 Appeal, at [201] & [207].

and Moggill, while remaining within the upper limit of non-damaging floods at Lowood (3,500m3/s).

3. The intent of Strategy W3 is to limit the flow in the Brisbane River at Moggill to less than 4000m3/s, noting that 4000m3/s at Moggill is the upper limit of non-damaging floods downstream.
4. The intent of Strategy W4 is to ensure the safety of the dam while limiting downstream impacts as much as possible.[144]

The NSW Court of Appeal summarising in these ways:

- [S]*ubject to preserving the integrity of the dam, the most significant consideration, which was reflected throughout the Manual, was the desirability of reducing the inundation of urbanised areas. That was because a larger number of people would be affected and greater damage would be suffered than by the inundation of less densely inhabited rural areas and rural infrastructure such as the low level downstream bridges*[145]; and
- *If the predicted maximum dam level exceeded 68.5m but not 74m, then there was sufficient risk that further inflows would result in urban inundation that the key objective became taking steps to prevent urban inundation. If the predicted maximum dam level were to exceed 74m, there would be sufficient risk of further inflows which may cause structural collapse to conclude that the key objective would be to preserve the dam. That meant taking steps which would probably cause hundreds of millions of dollars of damage to Ipswich and Brisbane, so as to minimise the catastrophic risk of a structural collapse of the dam (which would cause even greater loss).*[146]

Applying that context to the W1 – W4 strategies, '.... *the critical distinction between strategies W2 and W3 on the one hand and W4*

[144] Appeal, at [209].
[145] Appeal, at [231].
[146] Appeal, at [232].

on the other turned on the maximum release of 4000m3/s at Moggill. As explained [in the Manual] *in bold under strategy W3:*

> *The intent of Strategy W3 is to limit the flow in the Brisbane River at Moggill to less than 4000 m3/s, noting that 4000 m3/s at Moggill is the upper limit of non-damaging floods downstream.*[147]

In short compass, the arguments of Rodriguez, and accepted by the trial judge, were that the Wivenhoe Dam engineers were too reticent to implement Strategy W3, which should have occurred as early as 7 January 2011.[148] An argument bolstered by a finding at first instance, unchallenged before the NSW Court of Appeal, that '*the rain forecasts available on 7 January "either did demonstrate, or should have demonstrated... the strong likelihood, bordering on certainty as the day progressed, that the storage level of Wivenhoe Dam" would exceed 68.5m*'.[149]

Did the non-implementation of strategy as prescribed by the Manual cause Seqwater to have been negligent?

To answer this depends on which legal standard Seqwater's conduct fell to be determined by.

Sub Chapter 5.5 – The applicable standard by which to have judged Seqwater's conduct.

In the course of its ultimate dismissal of the class action, the NSW Court of Appeal postulated the question as whether the asserted acts and omissions of Seqwater in operating Wivenhoe Dam in early January 2011 were so unreasonable that no dam operator in Seqwater's position could properly consider them to be a reasonable exercise of that function.[150]

A yes answer to this question would have caused Seqwater to have lost the class action.

[147] Appeal, at [231].

[148] Appeal, at [532-533].

[149] Appeal, at [534].

[150] Appeal, at [656].

An important question then.

Which begs a question of its own: from where did that tightly formulated question emerge?

The answer was Section 36, as it then was, of Queensland's *Civil Liability Act 2003* ('the CLA'):

> 36 Proceedings against public or other authorities based on breach of statutory duty
>
> 1. This section applies to a proceeding that is based on an alleged wrongful exercise of or failure to exercise a function of a public or other authority.
> 2. For the purposes of the proceeding, an act or omission of the authority does not constitute a wrongful exercise or failure unless the act or omission was in the circumstances so unreasonable that no public or other authority having the functions of the authority in question could properly consider the act or omission to be a reasonable exercise of its functions.

Seqwater possessed requisite standing to seek the coverage of this section; the NSW Court of Appeal concluded Seqwater was

> *aptly described as a "public authority" because it is established under statute, has the functions and powers conferred by...statute, has no corporators or individuals who would benefit from the exercise of its powers as members of a corporate body, is responsible for the supply of water and other services relating to the water industry and is run by a board appointed by the responsible Ministers.*[151]

Therefore Seqwater's obligation to undertake its functions '*as a commercial enterprise*' and, presumably, charge for the supply of water and yield an appropriate profit, did not disqualify Seqwater from designation as a '*public authority*' as that was '*merely an application of the "user pays" principle, which is commonplace in the provision of public services*'.[152]

---

151 Appeal, at [117].

152 Appeal, at [118].

Seqwater was therefore entitled to argue that Section 36 of the CLA should apply to the facts asserted against it.

And why wouldn't it; even for non-lawyers, the plain words of the text of sub-section 2 provide a very substantial hurdle to clear.

Worsened by the fact that no evidence of such unreasonableness was presented at the trial. The NSW Court of Appeal noted the way the class action claimants' case was run at trial, concluding:

> [N]*o witness gave evidence of what a dam operator would or would not consider so unreasonable that what occurred could not properly be regarded as a reasonable exercise of that function.*[153]

Nor could the NSW Court of Appeal itself '*with no real understanding of the operation of a dam with water storage and flood mitigation functions... supply that gap in the evidence*'.[154]

Rodriguez '*attempted to address the gaps in evidence by submitting that any breaches by the flood engineers were "flagrant"*.'[155]

Had that finding been made the result could well have been very different.

However that '*pejorative*' was not available on the facts; the NSW Court of Appeal was not persuaded:

> *Use of that pejorative ordinarily conveys conscious breach of rights (such as a flagrant infringement of copyright) or, at least, a reckless indifference to the position. But Rodriguez has not established that the flood engineers were not attempting to take steps which they sincerely believed at the time to be properly directed to mitigate flooding. The contemporaneous documents do not suggest that the flood engineers were doing other than attempting to perform that function. If there were evidence, or cross-examination, to the effect that the flood engineers were deliberately taking steps to inundate urban areas, or failing to*

---

153 Appeal, at [656].

154 Appeal, at [657].

155 Appeal, at [658].

take steps which would avoid inundating urban areas, or were recklessly indifferent to that prospect, then this Court was not taken to it and could not in any event assess it given the limited scope of Rodriguez' contentions on the appeal'.[156]

Leading the NSW Court of Appeal to declare:

Effectively there were two contributing causes that led to urban inundation, that is to say flooding independent of any release from Wivenhoe and inflows into the dams on 11 January which required substantial releases from Wivenhoe. But on 7 January, and for most if not all of 9 January, it was not clear that there would be any urban inundation at all. It was certainly not clear that the issue confronting [Wivenhoe Dam engineer] *Mr Malone on 7 and 9 January was how best to minimise the urban inundation which (it is now with the benefit of hindsight known) was to occur on 12 and 13 January.*[157]

In fact,

the evidence suggested there had been a debate about whether to keep downstream flows to 3,500m3/s or 4,000m3/s. In particular, the trial judge referred to a note of a meeting at 08:30 on 10 January which "indicates that one of the topics was '3.5 and 4', which I infer was a discussion about the possibility of urban flooding resulting from an outflow rate of 3,500m3/s or 4000m3/s"....The fact that there appear to have been discussions about which outflow rate to choose suggests that this was not a decision which was so unreasonable that no dam operator could properly have considered the selection of the lower rate to be reasonable.[158]

Further evidence existed that, on 10 January 2011, four requests from the Brisbane City Council were recorded seeking to restrict flows at Moggill to 3,500m3/s.[159] This evidence confirmed for the

---

156 Ibid.

157 Appeal, at [686].

158 Appeal, at [622].

159 Appeal, at [623].

NSW Court of Appeal '*that active thought was being given to the* [Brisbane City] *Council's view that combined flow rates should be kept to 3,500m3/s, notwithstanding that operating in strategy W3, the Manual authorised flow rates that "should not exceed 4,000m3/s"*'.[160]

The Court of Appeal therefore concluded that '*Rodriguez could not discharge its onus in making out a contravention of the standard set by s 36(2)* [of the CLA] *in respect of any alleged breach. Seqwater's appeal with respect to the findings of breach on 6-10 January must be upheld. The finding of liability on the part of Seqwater must be set aside.*'[161]

In plain language, the trial court's finding that Seqwater and the two other defendants were liable in damages to all the 2011 flood victims, was wrong, and was therefore quashed.

Sub Chapter 5.6 – Ultimate determination of the effect of the Manual

The two preceding sub-chapters demonstrated that the Manual was not followed to the letter during 2011; for the reasons essayed immediately prior though, that was not fatal to Seqwater's defence.

The arbitrary nature of the Manual became apparent due to the exigencies of the crisis; the intent of the Manual's Strategy W3 was to limit the flow in the Brisbane River to less than the upper limit of non-damaging floods downstream flow rate of 4000m3/s at Moggill. In real time however, that upper limit failed; the NSW Court of Appeal's judgment cited '*a senior officer at Brisbane City Council*' having phoned Wivenhoe Dam's flood operations centre, called the FOC, at 00:45 on 10 January 2011 '*to advise that in the circumstances which actually prevailed, flow rates at Moggill above 3,500m3/s were damaging*'.[162]

Proving that, when taking into account on the ground downstream empirical effects,

---

[160] Appeal, at [624].
[161] Appeal, at [687].
[162] Appeal, at [627].

> [T]*he 4,000m3/s rate in the Manual is scarcely precise. The flood engineers were being told, by a person whose views might reasonably be regarded as carrying weight, that making releases which would result in a flow rate of 4,000m3/s at Moggill would cause millions of dollars of damage.*[163]

Moreover the Manual was '*a practical document addressed to engineers, not lawyers*', the NSW Court of Appeal noted, adding that

> *documents addressed to practical people skilled in the particular trade or industry*' ought '*be construed in light of practical considerations, rather than by a meticulous comparison of the language of their various provisions such as might be appropriate in construing sections of an act of Parliament.*[164]

Hence their Honours' appellate finding that '*the fact that the relevant acts and omissions may have involved a departure from the Manual, even in a serious way, does not of itself entail breach of the standard in s 36(2)*'[165] of the CLA. Bearing in mind, again, the heightened test under that section of so unreasonable that no dam operator in Seqwater's position could properly consider them to be a reasonable exercise of that function.

Sub Chapter 5.7 – Could the result not have been different had a different legal test applied?

Section 9 of the *Civil Liability Act* can best be described as the 'garden variety' duty of care that even non-lawyers know and comprehend. For example, the duty of care all drivers owe to other drivers on the road to keep a proper lookout and maintain control over their own motor vehicles so as to not cause collisions and other accidents.

Necessarily substantially easier to prove and apply to factual situations than a concept of conduct '*so unreasonable that no public or other authority ... could properly consider ... to be a reasonable exercise of its functions.*'

---

[163] Appeal, at [627].

[164] Appeal, at [224], [226] and [229].

[165] Appeal, at [650].

This was why Seqwater, having lost at first instance, urged the NSW Court of Appeal

> *the legal submission that the standard of care to be applied was not that of reasonable care under s 9 of the* [CLA], *but rather the attenuated standard required of public authorities under ... s 36 ... The primary judge found that s 36 ... was not engaged: as a result, the standard of care applied in judging Seqwater's conduct was the ordinary standard of a failure to take reasonable precautions against a foreseeable and not insignificant risk of harm ... However, if engaged, s 36(2) provided that an act or omission was not "wrongful" unless it was "so unreasonable that no public or other authority having the functions of the authority in question could properly consider the act or omission to be a reasonable exercise of its functions.*[166]

Rodriguez' legal argument before the NSW Court of Appeal, urging that Court not to apply that very high threshold standard, was that because Rodriguez' claim against Seqwater, SunWater and the State of Queensland was framed in common law negligence (that is that garden variety concept of negligence akin to what road users owe other road users, as opposed to a more confined and esoteric concept like breach of a statutory duty), Section 36 of the CLA was not engaged.[167]

That argument was not accepted on appeal:

> *The term "duty" is defined in the Dictionary to the* [CLA] *to mean a duty of care "in tort", "under contract", or another duty under statute or otherwise that is "concurrent with a duty of care in tort or under contract". The fact that the defined term "duty" is not used in s 36 suggests that the section was not limited to any particular cause of action.... Further, the phrase "breach of statutory duty" ... is at best a paraphrase of the third limb of the definition of duty, the meaning of which is not with-*

166 Appeal, at [12].

167 Appeal, at [65].

> *out its own difficulties. Finally, s 36(1) makes explicit provision for the application of the section, and does so by reference to "function" rather than the type of duty which is alleged to have been breached.*[168]

Concluding that the '*better view is that s 36 should not be read down as applicable only to a proceeding in which the cause of action is a breach of statutory duty*'.[169]

The other difficulty the text of Section 36 caused for Rodriguez was the broad scope of the word '*functions*' in its sub-section 1. Rodriguez argued successfully to the trial judge that none of the functions of Seqwater vested in it by legislation equated to '*functions*' in the Section 36 sense.[170] However the NSW Court of Appeal quite unambiguously disagreed:

> *On any view, Seqwater in fact undertook its flood mitigation functions by establishing a Flood Operations Centre and exercising its powers to release water in a controlled manner. In doing so it was, on Rodriguez' own case, subject to a duty to apply the ... Manual according to its terms. The Manual ... imposed strategies for the operation of Somerset and Wivenhoe Dams. It is not possible to read the reference to a "function" in s 36 as not encompassing the flood mitigation activities undertaken in January 2011.*[171]

Interesting then that the existence essayed in sub-chapters 5.4 – 5.6 of the Manual was one of the evidentiary bases which enabled Seqwater to argue, successfully, that it was a '*public authority*' which '*functioned*' for the purposes of Section 36 of the CLA. With all the attendant benefits that that section conferred.

Further successfully for Seqwater were the ways in which the Manual was construed by the NSW Court of Appeal and applied to

---

168 Appeal, at [68].

169 Appeal, at [69].

170 Appeal, at [74].

171 Appeal, at [82].

the operation of Wivenhoe Dam during those early January 2011 days. Leading to that Court's ultimate determination that, notwithstanding '*departure from the Manual, even in a serious way*', the high threshold test set by Section 36 was not met.

How bitterly ironic for the unsuccessful class action claimants then that, on the one hand, Seqwater was afforded by Section 36 of the CLA with a statutory defence with a much higher threshold test for its challengers to meet, so as to exonerate its '*departure from the Manual*' conduct. Yet, on the other hand, the very existence of that Manual entitled Seqwater to argue that it was a '*public authority*', so as to have enabled it to have taken advantage of Section 36 of the CLA and without which it might very likely have been unable to have done so.

Even after the NSW Court of Appeal delivered its verdict, and perhaps in light of the outsized utility Section 36 afforded Seqwater, Rodriguez maintained objection to the '*public authority*' characterisation of Seqwater at High Court of Australia level when seeking leave to appeal against the NSW Court of Appeal's dismissal of the class action:

> *The second error, we contend, is that, even in cases where section 36 may apply, it is limited to proceedings based on the alleged wrongful exercise – or failure to exercise – of a power or function of a statutory authority, that is a power or function being exercised in the capacity of a public authority. In the present case, the relevant power being exercised was the conduct of water operations which were authorised under licence and could be carried out by a public or a private body.*[172]

Seqwater's barrister argued, in response:

> [T]*he... proposition ... is that the section is somehow confined to breaches of duty ... directly conferred upon a public authority by statute and not enjoyed by other legal person... the obvious*

[172] *Rodriguez & Sons Pty Ltd v Queensland Bulk Water Supply Authority* t-as Seqwater & Ors [2022] HCATrans 61 (12 April 2022).

> *problem ...* [is] *that the applicant* [Rodriguez] *seeks to read words into the section that are simply not there. Even if they were there.... the respondent* [Seqwater] *was exercising a function directly conferred upon it by statute – namely the functions of flood prevention and flood water control ... as the Court of Appeal set out in some detail.*[173]

Reasons for decision were not rendered by Justices Keane, Steward & Gleeson who presided over the special leave application; after adjourning for no more than 10 minutes after addresses ceased, their Honours concluded:

> *The* [High] *Court does not consider that the case presents a suitable vehicle for consideration of the question of the operation of section 36 of the* [Civil Liability] *Act. That being the case, special leave is refused with costs.*[174]

In plain language, the class action officially ended then and there.

Given the absence of reasons for the High Court of Australia's decision, the competing arguments of the barristers are the most to go on as regards Seqwater's standing to have invoked the statutory advantage conferred by Section 36 of the CLA.

Richard Douglas KC, an expert insurance law barrister, hypothesised following publication of both the NSW Court of Appeal judgment and the High Court's dismissal of the application for leave to appeal against it this way:

> [T]*he gravamen of the appellate* [NSW Court of Appeal] *decision ... serves to underscore the ascendency of s*[ection] *36 as a powerful tool in the armoury of Queensland state and local government authorities ... as a defence in the hands of the State of Queensland, or a state or local government authority, as defendant.*[175]

---

[173] Ibid.

[174] Ibid.

[175] Richard Douglas KC, 'The Brisbane River Flood litigation falls – CLA s 36 rises' Hearsay', Bar Association of Qld, Edition 88, June 2022.

Rebecca Gilsenan from Maurice Blackburn Lawyers, the firm which ran the class action at all three court levels, was quoted after the High Court's dismissal along similar themes to Douglas KC, albeit viewed through an entirely different lens:

> *We think the* [NSW] *court of appeal made a mistake in lowering the standard of care that applies to Seqwater and other public authorities to a point where they're virtually immune from being held to be negligent.*[176]

There were other reasonings of the NSW Court of Appeal, which will be identified in the next sub-chapter below, which also contributed to Rodrieguez' ultimate loss. However Douglas KC and Gilsenan's commentaries that Section 36 of the CLA lowers standards of care of public authorities in such '*powerful*' ways, almost to a point of immunity, are both correct. The CLA was enacted by the parliament of Queensland in 2003 – similar analogues were enacted in other jurisdictions – in a rather frenzied context following the 2001 entry into liquidation of insurance giant HIH Insurance, which, with losses totalling $5.3 billion, remains Australia's largest corporate collapse even 22 years on.[177] Frenzied because of many and voluble arguments at the time of HIH's collapse seeking to blame a United States style plaintiff-friendly litigious culture for HIH's demise rather than that company's own internal failings. Of which there were many. The CLA nonetheless did demonstrably reform the law of negligence; a brief review of its contents says as much. Restrictions, either wholly or in part, of rights of claim exist to '*criminals*' and persons engaged in '*dangerous recreational activities*'. Reciprocally, '*protections*' were enacted for '*volunteers*', '*food donors*' and persons '*rendering first aid*'. Plus, of course, the enactment of Section 36, the section that vexed Rodriguez as far as the High Court of Australia, quite literally the highest court in the land.

---

[176] Marina Trajkovich, 'Thousands of Queenslanders lose class action lawsuit calling for greater 2011 flood compensation', Nine News Brisbane, 6:29pm, 12 April 2022.

[177] Paddy Manning, *Born to Rule – The Unauthorised biography of Malcolm Turnbull*, Melbourne University Press, 2015, p. 211.

Could the result of the class action have been different had the 2011 flood occurred instead during January 2001, before the CLA was enacted, and hence fell to be decided in a common law garden variety negligence context only?

Arguably yes, however even then that was not the way the NSW Court of Appeal ultimately viewed proceedings.

Sub Chapter 5.8 – Standards of care aside, was Wivenhoe Dam operated negligently in January 2011?

The class action claimants' argument against the logic of the independent contributions of the 48.416% of downstream sub-catchments propositions first published by myself and the NSW Court of Appeal some nine years later was that there should have been greater releases from Wivenhoe Dam on 6, 7, 8 and 9 January 2011. Had there have been, greater flood storage capacity would have remained in Wivenhoe Dam and correspondingly less need would have existed to have made the critical releases, which in fact combined (16 hours later) with the period of peak natural flows at Moggill[178] identified at the conclusion of sub-chapter 5.5.

Much debate therefore turned on whether or not Seqwater's engineers gave sufficient and appropriate weight to the 4-day forecasts provided on 7, 8 and 9 January 2011:

> ... *there was evidence that the flood engineers did in fact have regard to the 4-day forecasts prepared by the Bureau* [of Meteorology] ... *the proper course was to determine whether on the information available, and having regard to levels of uncertainty inherent in the forecasting, the steps taken by the flood engineers were reasonable in all the circumstances.*[179]

Rodriguez accepted that the Manual did not permit opening Wivenhoe Dam's gates in a flood event until its level rose to 67.25m; this stemmed from a '*command*' in section 8.3 of the Manual: '*The*

[178] Appeal, at [653].

[179] Appeal, at [306].

spillway gates are not to be opened for flood control purposes prior to the reservoir level exceeding EL 67.25'.[180]

Up to 11.00am on 5 January 2011 Wivenhoe Dam's level stayed below this threshold.

From 11.00am on 6 January 2011 however, Wivenhoe Dam's level crept above, and continued to rise. Slowly at first (6-9 January 2011), then rapidly, and unnervingly, after 11.00am on 9 January 2011:

11.00am on 2 January 2011	67.10m
11.00am on 3 January 2011	67.16m
11.00am on 4 January 2011	67.18m
11.00am on 5 January 2011	67.24m
11.00am on 6 January 2011	67.34m;
11.00am on 7 January 2011	67.81m
11.00am on 8 January 2011	68.59m
11.00am on 9 January 2011	68.54m
11.00am on 10 January 2011	71.95m
10.00am on 11 January 2011	74.10m.
12 January 2011	74.78m.[181]

11 January 2011 was described by Rob Ayre, Terry Malone and John Ruffini – respectively Senior Duty Flood Engineer, Duty Flood Engineer and Relief Senior Duty Flood Engineer on duty at the time – this way:

> *At 09:00 on ... 11 January 2011 ... Wivenhoe Dam was at 73.81m AHD (6.84m above FSL and only 0.19m below the W4 threshold level) and rising quickly ... Strategy W4, the dam safety strategy, was implemented at around 8:00 on ... 11 January 2011 in response to the intense rainfall in the immediate vicinity of the dam, which provided certainty that the lake level*

[180] Appeal, at [200].

[181] Appeal, at [301].

in Wivenhoe Dam would exceed the W4 threshold ... Gate releases from Wivenhoe Dam were then increased throughout the day ... the gates were all opened to a gate setting of 12m and the peak release was 7,460m3/s. Wivenhoe Dam lake level peaked at EL 74.97m AHD, which was 0.53m below the initiation threshold for the first fuse plug embankment'.[182]

Though the initiation level of the first fuse plug, as analysed in sub-chapter 5.4, is 75.7 metres, the invert of the pilot channel is set at 75.5 metres. In 2011 the dam engineers on duty used the invert level as the threshold, noting that it would take approximately 35 minutes for the first fuse plug to erode completely[183]: '*Until we have an event that triggers the fuse plug, we will not know for certain at what level the fuse plug will actually 'fail'* '.[184]

In retrospect the fear and terror which permeated the City of Brisbane was justified when one considers the rapid rise in Wivenhoe Dam height level from 68.54m to 74.1m at 10.00am on 11 January 2011 in a mere 47 hours. Then further still to the 74.97 metres witnessed by Ayre, Malone and Ruffini. As I recollected in my previous work, I arrived back to Brisbane from a New Zealand holiday on 11 January 2011 to a Central Business District in the throes of being evacuated; the fact that city law-firms, for whom the billable hour is the holy grail engrained in all of their lawyers, instead exhorted staff to leave early, spoke volumes. Little wonder '*Gosh this really is serious now*' was being tapped out by said firm employees on Facebook as they departed, and little wonder still that I typed my own update the next morning: '*Graceville...The suburb I live in and grew up in is rapidly submerging*'.[185]

That the relatively short distance left between the 74.97 metres recorded by Ayre, Malone and Ruffini and the Armageddon

---

[182] Ayre, Malone & Ruffini, pp. 8 & 9.

[183] Rob Ayre, interview with the writer.

[184] Ibid.

[185] *Tennyson Breach*, p. 78.

scenario of full dam collapse upon reaching a height of 80m was, notionally anyway, a mere two days away (based on an almost identical distance having been achieved during the 47 hours between 9 and 11 January 2011), only further served to heighten collective fears.

In fact, the 75.5m at which the first fuse plug would, theoretically, begin to start eroding towards failure and then send '*substantial flows of water in an uncontrolled way down three separate spillways*'[186] to Brisbane was even closer; a tiny in the overall scheme of things 53 centimetres.

Not even two primary school wooden student rulers away.

The sudden and sinister rise in Wivenhoe Dam's level during those frantic days is explicable by the rate of its inflows: '*Wivenhoe recorded an inflow at 08:00 on 10 January of 10,100m3/s and a peak inflow at 13:00 on 11 January at 11,600m3/s.*'[187]

Which in turn were explained by the sheer quantum of rain which fell in just nine hours alone on the most critical day for Wivenhoe Dam's operation – 11 January 2011 – to add to the rains delivered during the two preceding days. Justice Beech-Jones at first instance described the rain which fell over Wivenhoe Dam on 11 January 2011 as of '*biblical proportions*':

> *On Sunday, 9 January 2011, the heavens opened. Over that day and the following two days rainfall totals approximating 350mm to 400mm in depth were experienced in the catchment areas above the dams. The rainfall on 11 January 2011 in the area of Wivenhoe Dam was of biblical proportions.*[188]

His Honour went on to extract the following graph showing the hourly falls received at Wivenhoe Dam between 5.00am and 2.00pm that day[189]:

---

[186] Appeal, at [30].

[187] Appeal, at [337].

[188] *First instance*, at paragraph [8] of the judgment summary.

[189] *First instance*, Ch 7, at [374].

Hourly rainfall stations around Wivenhoe Dam reservoir

Hour ending

	Lowood	Savages Crossing	Wivenhoe Dam	Mt Glorious	Kluvers Lookout	Mt Mee	Somerset Dam	Caboonbah	Toogoolawah	Rosentretters	Cressbrook Dam
	6646	6559	6636	6680	6610	6690	6590	6574	6604	6553	6523
	mm	mm	mm	mm	mm	mm	mm	mm	mm	mm	mm
05:00 11 Jan	3	1	3	14	12	14	37	32	23	19	13
06:00 11 Jan	16	16	20	27	26	24	40	24	3	4	18
07:00 11 Jan	43	31	32	28	46	29	4	6	2	1	0
08:00 11 Jan	53	86	35	57	7	9	3	10	0	0	0
09:00 11 Jan	56	93	38	71	40	15	0	4	0	0	0
10:00 11 Jan	19	18	32	51	36	16	0	0	0	2	0
11:00 11 Jan	51	36	31	50	50	24	8	2	3	0	1
12:00 11 Jan	34	18	36	39	33	33	3	4	5	5	3
13:00 11 Jan	39	33	52	28	33	59	24	11	2	0	1
14:00 11 Jan	56	33	39	28	20	9	19	24	3	0	2

Perhaps however His Honour's epithet of '*biblical*' would have been better kept in reserve to be applied 11 years later so far as Mt Glorious was concerned, given that notwithstanding the 393mms which fell there in just nine hours on 11 January 2011 – contributing to a total of 656mm – 2022's 1.77 metre event total was almost three times that.

Thankfully the rainfalls and consequential inflows into Wivenhoe Dam ceased. The worst-case scenarios of fuse plug erosion sending '*substantial flows of water in an uncontrolled way down three separate spillways*' did not occur, though that was cold comfort to the tens of thousands of residents and business owners who endured the extensive Brisbane River flooding over the next two days, including the 13 January 2011 flood peak of 4.46m at the City Gauge.

Returning upstream to the operations centre, and to the forecasts available at the time,

> *... there was evidence that the flood engineers did in fact have regard to the 4-day forecasts ... the proper course was to determine whether ... the steps taken by the flood engineers were reasonable in all the circumstances.*[190]

---

[190] Appeal, at [306].

The '*circumstances*' including a dam height at 11.00am on 2 January 2011 of 67.10m, being 10cm greater than full supply level, as a result of substantial rains and consequential inflows necessitating 966,000ML of water having had to have been released from Wivenhoe Dam to the end of December 2010.[191] Such was the quantum of the December 2010 rainfalls and inflows that they themselves gave rise to a concept of '*December flood event*'[192] (even though what was received during December 2010 was trifling compared to what came not long after). The 2 January 2011 rise over full supply level by only 10cm meant that that particular quantum was sufficiently low enough for the Manual to have conferred a discretion:

- *The Manual conferred a discretion when dam levels were at 67.1m. It was open to the flood engineers to form the view, on 2 January 2011, that it was not necessary to return to FSL* [full supply level] *by 3 January 2011, and to shut the gates*[193]; and
- *The Manual contemplated that there are environmental benefits in water not being released through the floodgates, including dealing with stranded lung fish, and that in circumstances where the water level is within 50cm of FSL, the flood engineers had a discretion to delay releases.*[194]

In addition Seqwater's success in the appeal was assisted greatly by the NSW Court of Appeal's findings as regards weather forecasting; over the course of a 780 paragraph judgment the NSW Court of Appeal, at various parts, made the following exculpatory observations:

- The forecast available at 10:00 on 2 January 2011 was for less than 5mm of rain in the next 24 hours and less than 10mm forecast for the next four days, with most of that predicted to fall on the fourth day[195];

---

[191] Ayre, Malone & Ruffini, at 7.

[192] Appeal, at [235].

[193] Appeal, at [320].

[194] Appeal, at [457].

[195] Appeal, at [321].

- On 3 January 2011, there was little if any rain, the four day forecast predicted rain of up to 150mm on 6 January 2011 and the eight day forecast '*forecast no significant rain beyond that time*'[196];
- '*….on 2 January 2011 … the relevant weather forecasts at that time did not predict heavy rainfall which would lead to an expectation that FSL* [full supply level of Wivenhoe Dam] *would be exceeded in a practical sense. At least, it was not demonstrated that an engineer who held such a view was acting unreasonably*"[197]; and
- '*There was a favourable weather outlook on 2 January 2011; less than 5mm of rainfall was forecast for the 24 hours ending 09:00 Monday, and the 4-day … forecast 2-10mm*'.[198]

Hence the NSW Court of Appeal's conclusions, again at differing stages of the judgment, that:

- [A] *flood engineer was not* required *in those circumstances to continue the flood event, and to release water so that the level fell below 67m … on the afternoon of 2 January 2011 it was open to an engineer to determine that the December flood event had ended sometime after 16:00, when the next day's … forecast was available, and with dam levels at around 67.05m*[199]; and
- *… we do not agree that a flood engineer acted beyond the range of reasonable discretion in not making further substantial releases of Brisbane's drinking water when the outlook was as favourable as it was on 2 January 2011.*[200]

The second of these two recitations differs only by the invocation of the '*other*' purpose of Wivenhoe Dam: the necessity to store

[196] Appeal, at [439].
[197] Appeal, at [455].
[198] Appeal, at [457].
[199] Appeal, at [322], emphasis in original.
[200] Appeal, at [464].

'*Brisbane's drinking water*' cited at paragraph 162 of the NSW Court of Appeal's judgment.

With a result, as a matter of law, that '*preclude*[s] *a finding of negligence (even on the ordinary standard) on the part of the engineers in failing to continue to reduce the level of Wivenhoe ...*'[201]

This therefore answers the allegation that there should have been greater releases from Wivenhoe Dam on 6, 7, 8 and 9 January 2011 in the negative.

Moreover it did so on the 'ordinary' standard of negligence, irrespective of the super-imposed statutory defence inherent in Section 36 of the CLA.

A similar theme was contained in the final Queensland Floods Commission of Inquiry report:

> *... dam operators do not have the gift of foresight. A large flood is indistinguishable from a small flood when the first rain falls. Operators' ability to respond to flooding is hindered by the inaccuracy of rainfall forecasts and gauges, river level gauges and modelling. All that can be asked is that they act competently on the best information available to them and report faithfully what they have done.*[202]

Reliance in any way on the QFCI report was expressly disclaimed by His Honour Justice Beech-Jones when rendering judgment in the class action trial:

> *...the bulk of the evidence before the QFCI and its report were not tendered and therefore could not be considered by this Court. I have not read the QFCI report.*[203]

Nor was the QFCI report mentioned at all in the 780 paragraphs of the NSW Court of Appeal judgment.

These twin omissions were entirely appropriate because the tasks

---

201 Appeal, at [323].

202 QFCI Final Report, p. 438.

203 First instance, judgment summary at [4].

of and the procedures adopted by commissions of inquiry are quite different to, and much more lenient in evidentiary and procedural senses, than what transpires in courts of law. As Commissioner Holmes stated, the QFCI '*brief was not to seek out wrong-doers but ... to make recommendations for the improvement of preparation and planning for future floods and emergency response in natural disasters, as well as for any legislative change needed*'.[204]

Sub Chapter 5.9 – The downstream catchment effects in any event

Described this way by the judge at first instance His Honour Justice Beech-Jones in the judgment summary, the Manual '*designated a flow rate of 4000m3/s in the Brisbane River at Moggill as the threshold point at which homes and businesses downstream of the* [Somerset and Wivenhoe] *dams would commence to be flooded*'[205] {m3s being short hand for cubic metres per second}.

His Honour elaborated on this theme further in the judgment itself:

> *... the peak flow at Moggill was at around 1.00pm to 2.00pm on 12 January 2011 ... approximately 10700m3/s and that the flow without releases at around the same time was approximately 5400m3/s ...On these figures, outflows from Wivenhoe Dam contributed somewhere between 4200m3/s and 5300m3/s to a peak flow at Moggill on 12 January 2011 of between 10420m3/s and 10700m3/s.* [206]

His Honour's very analysis was cited by the Court of Appeal:

> *...even if there could have been no releases from Wivenhoe whatsoever, there would have been damaging floods ... slightly less than half of the peak flow at Moggill was attributable to releases from Wivenhoe ... Even if there were zero releases from Wivenhoe, there would have been urban inundation; the peak flows at Moggill without Wivenhoe exceeded 5,000m3/s*

[204] QFCI Final Report, p. 30.
[205] First instance, at paragraph [10] of the judgment summary.
[206] First instance, Chapter 7, at [403] – [404].

> *... natural flows at Moggill emanating from catchments below Wivenhoe, that is, without any allowance for water released from Wivenhoe.... reached a peak of some 5,800m3/s at about 00:00 on 13 January* [2011].[207]

Dare it be said again, a cognate hypothesis was penned by this writer some nine years before the NSW Court of Appeal did, albeit without the in-depth science and mathematics.

Referring back to the topic of flow effects, this image provided to the writer by Rob Ayre proves the truly awesome – in the original, non Hollywood, non Nickelodeon, sense in the way that word has since become erroneously construed – sight that is the releasing of Wivenhoe Dam waters on 12 January 2011:

The sunny day nature of the image is consistent with the engineers' recollection that '[B]*y around 15:00 on ... 11 January 2011 the heavy rainfall throughout the basin abruptly ceased*'.[208] Of course, the inflow impacts of the rains up to and including that time and date kept pushing the Dam level as high as 74.78 metres, a mere 19

---

[207] Appeal, at [652], [653], [684] & [688].

[208] Ayre, Malone & Ruffini, p. 9.

centimetres less than the 74.97 metre peak height the day before, notwithstanding the velocity of the releases depicted in the picture.

Keen-eyed viewers of the picture will note the observation deck to the left of the array of the radial gates looking downstream. A photo taken from that vantage point during October 2022 is included in Chapter 7 below. Suffice to say, the observation deck was well and truly closed to sightseers on 12 January 2011 for safety's sake. So was the Brisbane Valley Highway which, as demonstrated in sub-chapter 5.4, traverses the dam wall. The lack of visible traffic was no co-incidence; allowing cars to travel over it during those heady days would have been extreme folly, so the Highway was prudently closed at a number of locations both upstream and downstream of Wivenhoe Dam from 10 January 2011.[209]

Another pictorial comparison worth making is to sub-chapter 5.4's photograph of the semi-trailer travelling in an easterly direction over a Dam with a completely empty flood compartment. The palpable visibility of the Dam wall in June 2022 in such a context contrasts quite starkly with the comparatively tiny amount of the wall visible on 12 January 2011 due to the vast consumption of the flood mitigation compartment by the muddy waters fed so greatly into the Dam by the of '*biblical proportions*' rains of 11 January 2011 combined with the substantial rains which also fell in the two preceding days.

Indeed, despite the epic nature of the 12 January 2011 outflows depicted in the overhead picture, and further despite the nominal only levels of rain that had fallen in the 24 hours to 9.00am on 13 January 2011, Wivenhoe Dam's level had only dropped to 74.66 metres by that time and date.[210] Again this proved the continuing residual effect of the prior rainfalls in the overall catchment area continuing to recharge Wivenhoe Dam from upstream as it disgorged its load so spectacularly at the dam wall.

209 Rob Ayre, interview with the writer.

210 Ayre, Malone & Ruffini, p. 9.

Sub Chapter 5.10 – The High Court closes the final possible door to a victory for Rodriguez

The High Court of Australia, as the name suggests, is Australia's highest court; the final court in the court hierarchy over and surpassing all of the state and territory supreme courts and courts and tribunals within Commonwealth jurisdiction such as the Federal Court and the Administrative Appeals Tribunal.

There are no further appeals after the High Court; win or lose it's the end of the road.

Nor is there any automatic right to obtaining an appeal in the High Court; rather those wishing to appeal are required to seek formal leave to appeal to that court. Leave to appeal hearings customarily provide each party with 20 minutes only to either state their case for why their matter deserves the deliberation of the country's highest court or maintain that their opponent's does not.

On 12 April 2022 the parties did exactly that: Rodrigeuz, as leave seeker, had 20 minutes to explain the flood victims' reasons for seeking to quash the NSW Court of Appeal's appeal decision and substitute either a decision identical to the first instance victory by the High Court or alternatively orders by the High Court remitting the appeal hearing backwards to be re-heard by a differently constituted bench of the NSW Court of Appeal.

Seqwater had a reciprocal 20 minutes to argue that no error vitiated the NSW Court of Appeal's decision.

The fact that the special leave application was heard a mere 1 month and two weeks after the 2022 Brisbane flood event had no bearing; it was not even mentioned. Nor should it have been. That said, a sub-conscious form of book-end can be discerned: the January 2011 floods spawned the class action and the February 2022 flood event dovetailed neatly, albeit by pure co-incidence, with the final termination of the class action.

Final termination because the High Court, in the result, formally dismissed Rodriguez' special leave application. After 130 days of hearings during the trial process and a full 10 days of legal arguments before the NSW Court of Appeal – the vast majority of appeals usually never incur more than a single day for all sides' arguments – the High Court only needed 54 minutes to end the class action once and for all.

Reference was made in sub-chapter 5.7 to the High Court's non-willingness to countenance any form of either removal or dilution of the effective immunity Maurice Blackburn's Gilsenan opined that Section 36 of the CLA confers on public authorities. The High Court also heard arguments that the NSW Court of Appeal was incorrect to have also exculpated Seqwater on the conventional standard of care in negligence. Those arguments too were unable to persuade that court to entertain a grant of special leave to Rodriguez.

Short as those 54 minutes were, the end of the road was reached.

Class action litigant, and former Ipswich City Council member, Paul Tully said the High Court's decision was "*an absolute kick in the guts*". Frank Beaumont, a Goodna victim from both the 2011 and 2022 floods, who became the face, or perhaps the chest, of the class action when he was interviewed shirtless by the media cracking open a bottle of champagne the day Justice Beech-Jones delivered victory in the class action's trial, was also devastated. "*I think it is absolutely disgusting,*" Beaumont said. "*I had 30 feet of water go over my home, and I'll be lucky to get $10,000. It's not fair.*"[211]

The prospects then of a brand new class action lawsuit, with all of its millions of dollars of fees and expenses needing to be invested, arising from the 2022 flood event appear remote.

Sub Ch 5.11 – Money flows to the Rodrigeuz claimants

Sub-chapter 5.10 concluded with Frank Beaumont's 12 April 2022 statement: "*I think it is absolutely disgusting…I had 30 feet of water go over my home, and I'll be lucky to get $10,000. It's not fair.*"

---

[211] Marina Trajkovich, 'Thousands of Queenslanders lose class action lawsuit calling for greater 2011 flood compensation', *Nine News* Brisbane 6:29pm, 12 April 2022.

If Mr Beaumont thought he had suffered more than enough anguish by that date, worse still was to come. When Beaumont's interim share of the overall payout from the non-appealing parties SunWater Ltd and the State of Queensland arrived some three months later, it was a mere '*$797.67, despite his Goodna home sustaining $556,000 worth of damage*'.[212] $797.67 being Beaumont's half share split with his ex-wife, suggesting a total quantum of $1595.34, which nonetheless is a mere 0.2869% of his asserted damage total of $556,000.

The interim nature of the payment Beaumont and others within the almost 7000 strong cohort of claimants received was due to the total payout not being able to be distributed '*until all legal matters associated with the case were finalised*'.[213] Exactly what all of those '*associated*' legal matters were was not elaborated upon in the ABC's report.

The official position of Rodriguez' lawyers Maurice Blackburn was that the actual compensation payable to individual claimants '*will not be the full value of your assessed losses. Your losses will be adjusted to take account of the following factors:*

- The settlement is with Sunwater and the State of Queensland who were held to be responsible for half of the liability in the case.
- Like all settlements, it reflects a compromise on the full entitlement.
- The counterfactual flooding scenario: some claimants' losses will be reduced to reflect the fact that not all of the loss was caused by the operation of the dams. In some cases, properties would have flooded anyway, for example due to rain which fell downstream of the dams, or where flooding would have occurred at the property due to the elevation and topography of the property and surrounding land, and/or the way other rivers or waterways in the vicinity of the property flooded.

---

[212] Laura Lavelle, 'South-east Queensland's worst-hit 2011 flood victims receive nothing from class action', *ABC News* online, 6 July 2022.
[213] Ibid.

- Court-approved legal costs…
- *You and/or your insurer have agreed to pay a percentage of the compensation that you recover to the litigation funder, Omni Bridgeway. These costs include project costs and project management fees and funding commissions …*'[214]

On 4 May 2021, being a little over five months after the first instance verdict was handed down, another judge of the trial division of the NSW Supreme Court, Justice Adamson, published judgment approving the SunWater Ltd and State of Queensland settlement.[215] This timing coincided with the then scheduling of the appeal hearing for 12 days commencing on 17 May 2021 and concluding on 1 June 2021,[216] though in the result only 10 days were necessary. Part of Justice Adamson's reasoning for approving the settlement was that

> [I]*t is in the interests of justice that they* [the claimants] *obtain some recompense (albeit partial, given that Seqwater is not party to the settlement) now rather than later and that they not be subjected to the risk that the entitlement found by the trial judge will be removed on appeal.*[217]

Which removal, of course, did happen some four months later.

Rather presciently in the result, Justice Adamson added:

> *The vicissitudes of litigation are stressful and expensive for litigants and may lead to uncertain outcomes. Cases involving the law of torts may be particularly susceptible to different judicial minds taking different views on the same facts.*[218]

Settlements being settlements, many of the details of the SunWater Ltd and State of Queensland settlement were private and

214 https://www.mauriceblackburn.com.au/class-actions/settlement-payments/queensland-floods-class-action/payment-details/

215 *Rodriguez & Sons Pty Ltd v Queensland Bulk Water Supply Authority trading as Seqwater (No 28)* [2021] NSWSC 467 (4 May 2021) {"the settlement judgment"}.

216 The settlement judgment, at [4].

217 The settlement judgment, at [23].

218 Ibid.

confidential as between the Court and the respective sets of lawyers; Justice Adamson's settlement judgment was necessarily broad brush in nature.[219]

For this reason, the evidence presented by Maurice Blackburn of the proposed methodology by which the settlement sum was to be allocated amongst claimants is beyond the scope of this analysis. Suffice to say however that Justice Adamson was duly satisfied by that confidential evidence:

> *The fairness to group members has been established by the evidence adduced as to the method by which the settlement sum is to be allocated.*[220]

What could nonetheless be discerned was judicial acceptance that the proposed settlement not only involved the payment of a sum representing a reasonable compromise in relation to the claims for damages and costs, it also addressed the process of quantifying the claims of individual group members:

> *I have taken into account the anguish that can be caused to those who have suffered loss as a result of the negligence of others (as the trial judge found in the present case) when their entitlement to damages is either deferred for a lengthy period or at risk of being overturned on appeal. In those circumstances, injured parties may be disheartened by the processes of the administration of justice and may yearn for an early resolution ... It is in the interests of justice that they obtain some recompense (albeit partial, given that Seqwater is not party to the settlement) now rather than later ...*[221]

Not that there was any '*later*' as the effect of Seqwater's successful appeal was to have ended any prospect of a further fund of money coming into being.

---

219 The settlement judgment, at [17].

220 The settlement judgment, at [26].

221 The settlement judgment, at [22-23].

Actually the contrary was true given that the effect of the appeal – and subsequent confirmation of that result by the High Court – was to require that Seqwater's legal costs of all stages of the litigation be paid from the SunWater Ltd and State of Qld settlement sum.

Meaning even lesser a sized pool for all of the claimants to be funded from.

A point noted by Maurice Blackburn in its explanatory memorandum published between the NSW Court of Appeal and High Court decisions:

> *Before distributing money to claimants, Maurice Blackburn will deduct from the settlement sum court-approved legal costs for the conduct of the case from 2011 to date, including costs of the trial and trial preparation, appeal, settlement administration and any adverse legal costs which are ordered to be paid to Seqwater because of its success in the appeal … As at early 2022 … Seqwater has won its appeal … meaning that … Seqwater is not responsible for any losses caused by the flood. In addition, if Seqwater's victory is not overturned by the High Court of Australia, claimants will regrettably be required to pay a portion of Seqwater's legal costs from the settlement sum.*[222]

The firm added that its own costs and disbursements were considered by the NSW Supreme Court, in the context of evidence from an independent costs assessor:

> *After carefully considering this evidence, the court made orders approving the deduction of what it considered to be fair and reasonable … from the settlement amount … approximately 56% comprised Maurice Blackburn costs and 44% comprised amounts paid to third parties such as court costs, barristers fees and expert fees.*[223]

This was based upon Justice Adamson's statement, under a heading of '*Matters relating to costs and funding*': '*I consider the costs and*

[222] https://www.mauriceblackburn.com.au/class-actions/settlement-payments/queensland-floods-class-action/payment-details/

[223] Ibid.

fees to be paid from the settlement sum to be reasonable and commensurate with the work done and the risk undertaken'.[224]

The effect therefore of the settlement judgment was that '[T]*he group members will obtain, through the settlement, a fair proportion of the losses they suffered in the 2011 Queensland flood, in a much more timely way than if the losses needed to be determined by a judge or by a court-appointed referee*".[225]

It is evident from Frank Beaumont's comments some 14 months after the settlement judgment was published by Justice Adamson that Beaumont disagreed with the settlement judgment's '*fair proportion of the losses they suffered*' epithet. '[Maurice Blackburn] *advertises 'We fight for fair'. Is this fair?*', Beaumont rhetorically asked.[226]

Or at least Maurice Blackburn used to have '*we fight for fair*' as its tagline. '*Experience you can count on*' is that firm's new slogan. No suggestion is made that that change was in any way motivated by the outcome of the floods class action proceedings, or by anything else.

Maurice Blackburn's Rebecca Gilsenan responded to ABC queries about the quantification complaints of Beaumont and other claimants:

> *I wish that I could give everybody their full compensation, not half compensation to represent a partial victory — not adjusted by flood adjustment factors.*[227]

The final three words of Ms Gilsenan's quote dovetails with what Justice Adamson noted in the settlement judgment:

> *I am told that the assessment of damages, if individual claims are to be assessed separately, will take a period of years. The proceedings are complex and involve not only the interpretation of*

[224] The settlement judgment, at [27].

[225] The settlement judgment, at [29].

[226] Laura Lavelle, 'South-east Queensland's worst-hit 2011 flood victims receive nothing from class action', *ABC News* online, 6 July 2022.

[227] Lavelle, op. cit.

> *the manual used by the defendants in flood management ... but difficult assessments of counterfactuals to determine the causal effect of the defendants' negligence. The assessment of damages, if determined by traditional courtroom methods, will be both lengthy and expensive.*[228]

It can readily be inferred that the '*flood adjustment factors*' Gilsenan referred to were part of the evidence put before Justice Adamson based on her firm's dedicated '*Flood adjustment factor*' sub-page which, though presumably part of the suite of information deemed confidential for the purposes of the settlement judgment, is now, it is posited, quite substantially and publicly explained on the firm's website.[229]

Maurice Blackburn therein cite '*the Court ... approved ... Flood Adjustment Model. This modelling has been established by an expert who determined both the actual flood level, and the* "counterfactual" *flood level that would have occurred if the dams were operated differently*'.[230] The firm goes on to explain the way in which that model differentiates between the flooding that *actually happened* and the flooding that *would have happened* if the dams were operated differently.[231] A Flood Adjustment Factor calculation, between 0 and 1, is then introduced to adjust losses to ensure compensation is only being claimed for the damage caused by operation of the dams:

> *A factor of 1 means that all flood damage was due to the operation of the dams and the losses assessed will not be reduced. A factor of 0 means all of the flood damage would have occurred even if the dams were operated differently and the losses assessed will be reduced to zero. If the flood adjustment fac-*

---

[228] The settlement judgment, at [19].

[229] https://www.mauriceblackburn.com.au/class-actions/settlement-payments/queensland-floods-class-action/flood-adjustment-factor/

[230] https://www.mauriceblackburn.com.au/class-actions/settlement-payments/queensland-floods-class-action/flood-adjustment-factor/ - emphasis in original.

[231] Emphasis also included in original.

tor is somewhere between 0 and 1, the losses will be reduced to take into account that some *of the flood damage would have occurred even if the dams were operated differently.*[232]

The ways in which Maurice Blackburn applied or propose to apply the flood adjustment factor to individual claimants is not public record and hence beyond the scope of this analysis. That the interim payment of Beaumont and his ex-wife was only 0.2869% of his asserted damage total of $556,000 suggests however that Beaumont's final payment, if any, will not be much greater. Secondly it suggests that the Flood Adjustment Factor so far as Beaumont's house was concerned had to have been much closer to 0 than 1 due to the inherently flood ravaged nature of Beaumont's suburb of Goodna overall. This is based on the explanation extracted immediately above that a factor of 0 applied to a specific property means all flood damage would have occurred even if the dams were operated differently, and the losses assessed reduced to zero accordingly. 31 Goodna properties were declared eligible as of August 2022 for voluntary buy-back based on what Ipswich Mayor Teresa Harding described as Queensland Government deeming that suburb '*an initial priority location … because of the extent of the disaster impact and flood risk*'.[233] A deeming such as this is inherently consistent with a Flood Adjustment Factor for houses in Goodna overall being much closer to 0 than 1. With consequentially low, perhaps even zero, monetary results for its residents who claimed in the class action.

A painfully ironic outcome given the high levels of early enthusiasm for investigating a class action lawsuit from that suburb; as of April 2012:

Maurice Blackburn … received an 'incredible' response to public meetings in Goodna (attended by 400), Chelmer (300), Ips-

[232] https://www.mauriceblackburn.com.au/class-actions/settlement-payments/queensland-floods-class-action/flood-adjustment-factor/ - emphasis in original.

[233] Jack McKay, 'Flood areas to go green', *The Courier-Mail*, 13 August 2022, p. 11.

wich (200), Fernvale and Wacol (each 100).[234]

That same edition of Brisbane Legal coincidentally reported the launch of your present writer's *Tennyson Breach* this way:

> *Topp's book ... also highlighted a key aspect of the 2011 floods that may be an issue for some of those now signing up for a possible class-action lawsuit against the Wivenhoe Dam operator Seqwater ... that flood inflows to the Brisbane River from the Bremer River and Lockyer Creek were downstream of Wivenhoe Dam and therefore could not have been stopped from surging towards Brisbane.*[235]

From '*may be an issue*' to, as sub-chapter 5.9 analysed, a contributing issue to the ultimate failure of the class action.

---

[234] *Brisbane Legal magazine*, April 2012, 'Flood challenges loom before class action begins', p. 4.

[235] *Brisbane Legal magazine*, April 2012, 'Writing with a river view', p. 14.

6

Wivenhoe Dam Management in 2022

The real factual contrast between 2011 and 2022 so far as pre-event total rainfalls were concerned meant that Wivenhoe Dam loomed less large in 2022.

At time of publication, no formal commission of inquiry akin to the 2011-2012 Holmes Commission of Inquiry had been established, nor any class action mooted.

A very substantial reason for why not being the enormous 'head start' Wivenhoe Dam and its engineers had in 2022.

Unlike the large volumes of rain which fell and ultimately saturated the Wivenhoe catchment in 2010 before the *'biblical proportions'* of rain in early January 2011, Chapter 1 essayed the way in which the 2022 flood event was preceded by a prolonged drought with a consequential inexorable downward trend in Wivenhoe Dam's drinking water capacity between 2015 and the end of 2021; its January 2021 capacity level fell to a worryingly low 36.3%, as vividly depicted in the Bureau's graph extracted in Chapter 1.[236] So much so that purified recycled water was meant to have been pumped into Wivenhoe Dam based on a trigger point level of 40% having been breached. Yet, for reasons not reported, perhaps because subsequent events of February 2022 rapidly and spectacularly reversed that pretext, no steps were taken to have augmented the Dam's then worryingly low capacity by supplies of purified recycled water.[237]

Hence this writer's invocation of a 'head start' analogy; Wivenhoe Dam's pre-flood event capacity in 2022 was much lower than it was in 2011 when heavy rains and consequential inflows during

[236] Supra, fn 8.

[237] Hayden Johnson, 'Billion Dollar Water Torture', *The Sunday Mail*, 14 August 2022, p. 9.

December 2010 created a concept of '*December flood event*' because, by 2 January 2011, Wivenhoe Dam's level had consequentially risen to 67.1m, 10cm above full supply level.[238]

Not only that, the Wivenhoe and Somerset catchments, as a result of the December 2010 rains, were '*effectively primed to generate runoff*'.[239]

No such similarity existed during January and February 2022. Rather the exact opposite was the case. So remarked by a spokesperson for Seqwater:

> *There was no need to make any early releases from Wivenhoe Dam. Prior to this event, on Thursday* [24 February 2022], *Wivenhoe had approximately 43 percent of its drinking water supply storage available and 100 per cent of its flood mitigation capacity.*[240]

4 days later, on 28 February 2022, Seqwater spokesperson Mike Foster confirmed that Wivenhoe Dam was at half its capacity when the 2022 event began on 23 February 2022, meaning half of its drinking water storage and an empty flood compartment created '*a very large space that we had available to manage this event*'.[241]

Clearly that was for the best given the mind-boggling rapidity of Wivenhoe Dam's upward capacity change:

> *Wivenhoe Dam capacity rose from 58.7 per cent on February 24* [2022] *to 183.9 per cent by February 27. That's a 125.2 per cent increase in dam capacity in three days and just shy of the 190.3 per cent total reached in … 2011…*[242]

This represented an increase of more than 1.4 million megalitres,[243]

---

238 Appeal, at [235].

239 Ayre, Malone & Ruffini, at 7.

240 Lydia Lynch, 'Dam overflow stepped up as rains return', *The Australian*, 3 March 2022, p. 5.

241 Hayden Johnson, 'One dam huge problem', *The Courier-Mail*, 1 March 2022, p. 5.

242 Sarah Richards, 'A break down of how south-east Queensland's flood crisis played out', *ABC News* Online, 21 March 2022/Lord Mayor Adrian Schrinner via LinkedIn, 6 September 2022.

243 Bureau of Meteorology, 'Special Climate Statement 76 – Extreme rainfall and flooding in south-eastern Queensland and eastern New South Wales', 25 May 2022, p. 14.

or 2.8 Sydney Harbour's worth of waters {based on 500,000 megalitres}, in an astonishingly short time. Over the course of the four days 25 – 28 February 2022, '*Wivenhoe and Somerset dams have collectively held back at least 2.2m megalitres of water, while 150,000 megalitres has been released*'.[244]

Put another way, Wivenhoe Dam's maximum 2022 dam wall height was 74.61 metres,[245] not materially lower than the maximum height of 74.97 metres witnessed by Ayre, Malone & Ruffini on 12 January 2011[246]:

> *Therefore, there was only approximately 60,000 megalitres of additional storage between 74.61 and 75 metres AHD (or approximately 5 per cent of the total volume of the Flood Storage Compartment reserved for flood mitigation purposes). The dam operators utilised almost the entire flood storage compartment, which delivered a close to optimal flood mitigation benefit to residents downstream.*[247]

The release, six weeks later, of the Office of the Inspector-General Emergency Management report from which the preceding quote was extracted, gave rise to a not unreasonable in the circumstances 13 October 2022 headline of '*Dam disaster averted*':

> *WIVENHOE Dam was just 39cm or 60,000 mega litres of water away from having its gates flung open ... "dam safety strategy" protocols ... focus mainly on ensuring the dam doesn't fail and "requires" gates at the tip to be fully opened. This critical water level is defined as 75m ... Another 60,000 mega litres of water would have been too much. That's only four percent of the total 1.46 million mega litres that had flowed into Wivenhoe in the previous three days.*[248]

---

244 Hayden Johnson, 'One dam huge problem', *The Courier-Mail*, 1 March 2022, p. 4.

245 Office of the Inspector-General Emergency Management, 'South-East Queensland Rainfall and Flooding February to March 2022 Review', 31 August 2022, p. 108 of 155.

246 Supra, fn 182.

247 Office of the Inspector-General Emergency Management, op. cit.

248 Madura McCormack and Stephanie Bennett, 'Dam Disaster Averted', *The Courier-Mail*, 13 October 2022, p. 10.

The reference in the 13 October 2022 article to '*requires*' brings back into focus the Manual, so exhaustively analysed during the course of the class action litigation reported on in Chapter 5. As matters transpired, the Manual was revised soon after the 2011 floods. Nine times in fact. Whether or not any of these revisions occurred precisely as a result of the QFCI is beyond the scope of this analysis, other than to note the final QFCI report's recommendations included:

> 17.6 The Queensland Government should ensure that all flood mitigation manuals include the requirement that those operating the dam during flood events hold current registrations as professional engineers.
>
> 17.7 Seqwater should consider engaging a technical writer to develop completely new manuals after the operational strategies for Wivenhoe, Somerset and North Pine dams are set by the Queensland Government.
>
> 17.8 Seqwater should ensure a legal review of the Wivenhoe manual and the North Pine manual is completed before the manual is submitted for approval
>
> ....
>
> 17.11 The assessment of flood mitigation manuals should be completed by a person with appropriate expertise who has had no involvement in its development, at any stage, and who can be seen to be independent of all individuals who were so involved.[249]

As an aside, the NSW Court of Appeal formed its own views of the Manual in its capacity as a document, which views were not-inconsistent with the thinking behind the QFCI's recommendations 17.7, 17.8 and 17.11:

> *The Manual was replete with grammatical and syntactical glitches. Capitalisation was haphazard ... That is not said as a*

[249] QFCI Final Report, p. 27.

> *significant criticism. The drafting was by engineers, who were evidently and understandably much more focused upon its substance than its form ... the Manual was, as Seqwater submitted and Rodriguez did not deny, "a practical document addressed to engineers, not lawyers". It falls within the principle that "documents addressed to practical people skilled in the particular trade or industry" ought "to be construed in light of practical considerations, rather than by a meticulous comparison of the language of their various provisions such as might be appropriate in construing sections of an act of Parliament".*[250]

It is fair to say that the current version of the Manual, available for free download from Seqwater's website, is a very polished document. The shortcomings and syntax awkwardness the NSW Court of Appeal noted have been well and truly ironed out in what is Revision 16 of the Manual, released November 2021. Revision 16 of the Manual quotes three broad '*Strategies*' designed to implement the four new '*Objectives*' for flood operations at Wivenhoe and Somerset Dams. New because they replace the former five objectives that were extant as of the 2011 flood event, namely:

- Ensure the structural safety of the Somerset and Wivenhoe dams;
- Provide optimum protection of urbanised areas from inundation;
- Minimise disruption to rural life in the valleys of the Brisbane and Stanley Rivers;
- Retain the storage at Full Supply Level at the conclusion of the Flood Event; and
- Minimise impacts to riparian flora and fauna during the drain down phase of the Flood Event.[251]

The former five have become the following 4:

---

250 Appeal, at [180] and [224].

251 Appeal, at [182].

Primary: Prevent structural failure of the Dams;

Secondary: Ensure the Water Supply Compartment of each Dam is full at completion of a Flood Event;

Lower Order: Mitigate downstream flooding; and

Lower Order: Protect the riverine and riparian environment.[252]

The three broad '*Strategies*' designed to implement these four new '*Objectives*' are, in temporal order:

- '*Flood Mitigation Strategy*';
- '*Dam Safety Strategy*'; and
- '*The Drain Down Strategy*'.

The '*Flood Mitigation Strategy*' '*only*' applies to situations where Wivenhoe's predicted level will be below 75 metres. Whereas above that level, a '*Dam Safety Strategy*' supplants the '*Flood Mitigation Strategy*'. The '*Dam Safety Strategy*' is devoted to preventing structural failure of the Dam by increasing releases '*and hav*[ing] *radial gates* ***fully open*** *before the lowest fuse plug embankment is breached*'.[253]

As stated in sub-chapter 5.8, 75.5 metres was the dam wall height level at which the first Wivenhoe Dam fuse plug[254] would, theoretically, start to begin eroding, and, upon failing some approximate 35 minutes later, send '*substantial flows of water in an uncontrolled way down three separate spillways*'[255] to Brisbane.

Interestingly the word '*theoretically*' adopted by the NSW Court of Appeal is the operative word because, since their 2005 installation, none of the Wivenhoe Dam fuse plugs have triggered. Meaning that, until they do, their modelling is exactly that. The current version of the Manual says as much:

It is important to note that the precise Lake Level at which the fuse plug embankment will breach is uncertain due to likely

---

252 Revision 16 of the Manual, released November 2021, p. 46.

253 Revision 16 of the Manual, released November 2021, p. 49 – bold font added. A similar theme is contained on page 53 thereof.

254 Supra, fn 183 &184.

255 Appeal, at [30].

drawdown effects in the upstream approach channel and the depth of overtopping required to erode the embankment materials. Professional judgment is needed when operating around these levels and visual monitoring is recommended.[256]

At least we can be certain that the Wivenhoe Dam 2011 and 2022 respective maximum wall heights of 74.97m and 74.61 metres were sufficiently low enough not to trigger the first fuse plug.

However nothing is theoretical about the sheer bulk of waters which would flow out of Wivenhoe Dam if its gates are required to be fully opened at 75 metres. '[S]*ubstantial flows of water in an uncontrolled* way' if the gates are not intentionally opened at 75 metres and, instead, begin to commence a breach of the first fuse plug at 75.5 metres would be barely less substantial in a controlled gate opening context given the way in which the Dam would be so heavily overloaded with water at levels of 75.0, 75.5 and 75.7 metres; refer again to Sub-Chapter 5.9's 12 January 2011 overhead picture to see just how swollen Wivenhoe was at levels not even as high as 75.0 metres. Full opening of Wivenhoe Dam's gates during February 2022, with commensurately extreme outflows of water, would have interacted with a city of Brisbane already, for the reasons stated and pictured in Chapter 2, overflowing with 'rain bomb' flooding of its own. Hence why the epithet in McCormack and Bennett's headline of '*disaster*' was fully warranted. Catastrophe would have been even more suitable a descriptor actually, bearing in mind the comparative lack of adverse affection by rainfalls in Brisbane city and surrounds themselves during January 2011.

Sure, persons and businesses who actually suffered loss and damage during the February 2022 flood event would not agree that their own personal disasters were '*averted*', to quote the third word of those journalists' headline. However looking at the matter dispassionately does enable the '*disaster averted*' comment to be validly applied.

---

256 Revision 16 of the Manual, released November 2021, p. 25.

Of course, the one fact which can not be avoided is, as postulated earlier in this chapter, the 'head start' that Wivenhoe Dam had in 2022 relative to 2011, due to the long history of drought in the lead up to the 2022 event. Readers of this work who have not already downloaded the Manual from Seqwater's website ought do so simply to look at the cover photo, which is an overhead of the Wivenhoe Dam wall, surrounded by palpably dry and parched grass sporting a sickly brown colour. Completely the opposite of the lush greenery that abutted the Dam's surrounds during this writer's field trip there during June 2022. Bearing in mind that Wivenhoe was '*just 39cm or 60,000 mega litres of water away from having its gates flung open*' in a context of that quantity of water being a mere '*four percent of the total 1.46 mega litres that had flowed into Wivenhoe in the previous three days*',[257] it ought be strikingly obvious that had the pre-2022 drought phase been only slightly lesser in its intensity, with a commensurately slightly higher starting level of waters, the resulting loss and damage for Brisbane overall would have been greatly worse.

Seqwater's Foster went on to describe the way in which Seqwater '*slowly but surely put controlled releases into the system and allowed us not to exacerbate those flows downstream*'.[258] Of those releases, the first began for 1 hour on the night of Friday 25 February 2022:

> *There were some releases…during the course of last evening … they were done at the time to do some strategic releases in order to manage water flow levels between the various parts of the SEQ network…Future releases will be looked at through the lens of the flood manuals …*[259]

A tweet by Seqwater on 26 February 2022 confirmed:

---

[257] Madura McCormack and Stephanie Bennett, 'Dam Disaster Averted', *The Courier-Mail*, 13 October 2022, p. 10.

[258] Hayden Johnson, 'One dam huge problem', *The Courier-Mail*, 1 March 2022, p. 5.

[259] Jeremy Pierce, Danielle Buckley, 'SEQ's deluge turns deadly', *The Sunday Mail*, 27 February 2022, p. 3.

> [G]*ated releases from Wivenhoe ... commenced temporarily at 10pm Fri 25/2 at a low rate and then ceased at 11pm due to downstream flooding ...There is still more than 90% of the ... flood storage compartment available.*[260]

We now know from the Office of the Inspector-General Emergency Management report that that '*90%*' of the flood storage compartment's availability reduced to a mere 4% extremely rapidly, so immense were the inflows into Wivenhoe Dam after 10pm on Friday 25 February 2022.

No further releases from Wivenhoe Dam, save for that 1 hour on the night of Friday 25 February, occurred until 4.00am on Sunday 27 February 2022. '*There's no planned release today*' being quoted by Queensland's Minister for Police & Emergency Services Mark Ryan during a media briefing at 8.45am on Saturday 26 February 2022, followed up by '[T]*here is no concern for alarm*' by the Premier herself at 1.45pm that afternoon.[261] It was not until later that day, as the intense rain fall continued, that '*Seqwater's latest modelling shows the recommencement of releases from Wivenhoe Dam is highly likely to occur overnight and into the early hours of Sunday* [27 February 2022] *morning. This situation is constantly evolving and we are responding rapidly to changing circumstances*'.[262] A follow up same-day tweet read '*Seqwater is continuing to plan for gated releases at Wivenhoe Dam in the early hours of Sunday morning. Releases from Wivenhoe take 16-24 hours to reach Brisbane. The high tide occurring in Brisbane at 7am tomorrow morning will therefore not be affected by these releases*'.

As it happened, though not disclosed on social or other forms of media at the time, an '*alternative operating procedure*' invoked at approximately 4:00pm on Sunday 27 February 2022 whereby there was a temporary reduction of outflows from Somerset Dam into Wivenhoe Dam '*designed to*' assist to '*slightly reduce the peak level in*

---

[260] Twitter @Seqwater, 26 February 2022.

[261] Ben Murphy, *7 News* Queensland telecast, Wednesday, 27 April 2022.

[262] Twitter @Seqwater, 26 February 2022.

Wivenhoe in combination with avoiding or minimising a need to increase outflows from Wivenhoe Dam into the Brisbane River'. Doing so meant that '*the threshold for transitioning from Flood Mitigation Strategy to Dam Safety Strategy ... was not met. Invoking the Dam Safety Strategy would have required increased emphasis to be placed on the primary objective in the flood mitigation manual of preventing structural failure of the dam*'.[263] Described in plainer language by a sub-headline to McCormack and Bennett's article as '*Engineers devise hack to save Wivenhoe*', '*engineers, using an "alternative" protocol, managed to prevent Wivenhoe Dam from hitting critical water levels which would have shifted ... focus ... to preventing the dam from failing*.'[264]

No surprise then that the Office of the Inspector-General Emergency Management, a body tasked to '*regularly review and assess the effectiveness of disaster management by the State* [of Queensland]',[265] declared some six months after the event, that '*dam operators ... delivered a close to optimal flood mitigation benefit to residents downstream*'.[266]

Sunday 27 February 2022 was an extremely rainy day in and around Brisbane – recall the images in Chapter 2 of the gigantic volumes of water Kedron Brook was carrying. Of course, Kedron Brook was wholly independent of the Brisbane River; as for suburbs which were not, damage had already been occasioned to those many Brisbane suburbs by Brisbane River flooding during either Saturday 26 or Sunday 27 February 2022 independently of Wivenhoe Dam releases. Independently because Wivenhoe Dam releases, which began at 4am on the morning of Sunday 27 February 2022,[267]

---

[263] Office of the Inspector-General Emergency Management, 'South-East Queensland Rainfall and Flooding February to March 2022 Review',31 August 2022, p. 107 of 155.

[264] Madura McCormack and Stephanie Bennett, 'Dam Disaster Averted', *The Courier-Mail*, 13 October 2022, p. 10.

[265] Office of the Inspector-General Emergency Management, 'South-East Queensland Rainfall and Flooding February to March 2022 Review',31 August 2022, p. 112 of 155.

[266] Ibid., p. 108 of 155.

[267] Twitter @Seqwater 27 February 2022.

had a 16-24 hour conveyance time. As further proven by the fact that the 3.85m peak City Gauge flood height of the Brisbane River was recorded as late into the event as the morning of Monday 28 February 2022, not earlier.

Frank Beaumont, cited at sub-chapter 5.10 after the High Court finally dismissed the class action of which he was the human face and sub-chapter 5.11 after his minuscule interim damages payment was finally received, was evacuated from his Goodna home as early into the 2022 event as the morning of Saturday 26 February 2022, declaring that it's '*2011 all over again*'.[268] The night before, specifically at 10.45pm on Friday 25 February 2022, the Ipswich City Council had issued prepare to leave warnings for low lying areas along Bundamba, Warrill and Woogaroo Creeks,[269] Woogaroo Creek being the creek which rises below Camira at an elevation of 30.6 metres and drops 24.3 of those metres along its 5.51 kilometre length to its Goodna junction point with the Brisbane River, meaning that, in practical terms, the suburb of Goodna is, just as 2011 proved, arguably South-East Queensland's most flood prone suburb.

What difference could however be discerned during 2022 was that Beaumont's local Ipswich city councillor and fellow class action litigant Paul Tully noted that Goodna waters were slowly rising '*bit by bit every hour*' as opposed to the '*came in a rush*' nature of 2011.[270]

Official graphics (following page) issued nine months later by the Ipswich City Council bear this out; at the official Ipswich City gauge on the Bremer River which transits the Ipswich CBD the gradual nature of the rise from 25 February to the 27 February date of Beaumont's Goodna evacuation is well demonstrated[271]:

---

[268] Hayden Johnson, 'Dam anxiety spills over', *The Courier-Mail*, 28 February 2022, p. 5.

[269] Ipswich City Council Media Statement, 'Prepare to leave warning issued as flash flooding to occur in low lying Ipswich areas', 26 February 2022.

[270] Johnson, op. cit., p. 4.

[271] Ipswich City Council 2022 flood review, p. 10.

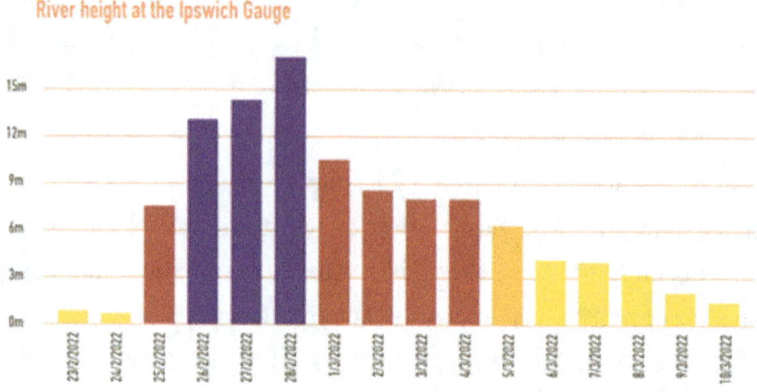

The comparative difference eye-witnessed by Tully infers a more natural rate of rising water levels due to actual from the sky rainfall rather than the necessarily artificially inflated surge rate of water rise due to intentional releases from Wivenhoe Dam. Releases which in 2022 were, as outlined earlier, quite deliberately low during the height of the crisis due the lack of necessity to invoke the Dam Safety Strategy with associated compulsory requirement to '*hav*[e] *its gates flung open*'. Karalee, being the suburb at the junction of the Bremer and Brisbane Rivers, and hence in close proximity to Goodna, received 695mm in the seven days up to 11am Monday 28 February 2022.[272] Of those seven days, only 31 hours existed between the Wivenhoe Dam release start time of 4.00am Sunday 27 February and 11am Monday 28 February, proving both the correctness of Tully's observation and the very low, if any, impact caused to Goodna by Wivenhoe Dam releases, save perhaps only for the final 31 hours of that week.

Closer to Brisbane's CBD is the inner-western suburb of Rosalie, which, just like Goodna, was adversely affected in both 2011 and 2022: '*Beck St Rosalie, awash once again in flood waters ... of last Sunday's big wet ... things went from bad to worse in Beck St at about 3.30pm*' as '*the waters flowing down Beck St went from "knee high to thigh high in minutes"*'.[273]

[272] Supra, fn 13.

[273] Frances Whiting, 'Beck Street Battles Back', *The Courier-Mail*, 5 March 2022, p. 5.

The '*last Sunday*' referred to being Sunday 27 February 2022.

An awful outcome for the people of Rosalie, whose suburb was devastated twice in 11 years. However the floodwater rises that necessitated Beaumont's Goodna evacuation and the Rosalie damage Whiting's many interviewees reported on the afternoon of Sunday 27 February had eventuated before any Wivenhoe Dam releases, save for the solitary 1 hour on the night of 25 February, could have had an impact.

Rather, and as *The Courier-Mail* correctly editorialised, '[T]*o flippantly say this is not as bad as 2011 is to ignore one critical point: that this flood has seen historic flows in pretty well every river and creek across the South-East*'.[274]

Including therefore the Brisbane River's sub-catchments downstream of Wivenhoe Dam which were not and could not have been held back by its dam wall.

And also those wholly independent of both, such as Kedron Brook.

The third and final of the Manual's '*strategies*' outlined in this chapter – the '*Drain Down Strategy*' – is, as its name suggests, purposed to empty Wivenhoe Dam's flood-storage compartment '*to minimise the likelihood of structural failure of the Dam*' by ensuring that that compartment is fully available should a new flood event occur at the expiry of the last one.[275] A very necessary strategy when one considers that that possibility was factually proven on 2 January 2011 when the '*December* [2010] *flood event*' ceased, only to be replaced so very soon after by the '*biblical proportions*' of rain which fell 1 week later.

This '*must*' stipulation therefore caused planned controlled releases over the ensuing week Monday 1-8 March 2022 so as to restore Wivenhoe Dam's flood compartment to zero percent capacity. Seqwater's Mike Foster explained that '*Mother Nature can change*

[274] Editorial, *The Courier-Mail*. 28 February 2022.

[275] Revision 16 of the Manual, released November 2021, p. 47.

really quickly, so if we were to get more rain coming in next week ... we need to ensure that the dam is ready to be able to deal with another event'.[276]

The plan therefore being to fully empty the flood compartment '*and bring Wivenhoe back to its full* [drinking water] *supply capacity*'.[277]

This concept of course recognising the ways in which flood mitigation, along with water supply, have been Wivenhoe Dam's two primary purposes. Yet, as the NSW Court of Appeal concluded, these two purposes '*are opposed to each other...*' and hence '*there must be some mechanism for regulating the compromise between*' those two purposes.[278]

In 2022 the '*compromise*' was as expressed above, including the way in which Seqwater's Foster was '*confident*' enough to have been quoted on 27 February 2022 that '*Wivenhoe Dam's floodwater storage compartment could be used without the need for a prolonged release*'.[279]

That confidence was well-placed, and the '*compromise*' worked, due to the rain, finally, easing off on 28 February 2022; on 13 March 2022 when all of the both metaphorical and literal water had run under many bridges, Seqwater's CEO Neil Brennan concluded:

> *Wivenhoe ... is a dual purpose dam – it's used for drinking water as well as providing flood mitigation ... the dam provides about 50 percent of South-East Queensland's drinking water supply. Prior to ... February 24 ... storage was at about half of its drinking water capacity and had 100 per cent flood mitigation capacity available. There was no need to make releases. Wivenhoe did its job ...* [it] *held back more than 2.2 million megalitres of water – the equivalent of four Sydney Harbours from Brisbane. The flooding that occurred in Brisbane was the*

---

[276] Hayden Johnson, 'One dam huge problem', *The Courier-Mail*, 1 March 2022, p. 4.

[277] Ibid.,

[278] Appeal, at [164].

[279] Hayden Johnson, 'Dam anxiety spills over', *The Courier-Mail*, 28 February 2022, p. 5.

result of rainfall impacting creeks and rivers downstream ... Seqwater made steady controlled releases, below the peak river levels so as not to exacerbate the already naturally occurring flooding. Quite simply, the releases made from Wivenhoe Dam did not contribute to the flooding that took place during this event.[280]

Fortuitously very little rainfall fell during the first week of March 2022, which enabled, between the final weekend of February and Wednesday, 2 March 2022, 700,000 megalitres[281] to have been drained from Wivenhoe's 2.08m megalitre flood-storage compartment. Notable in this regard being that the 1750 cubic metres of water per second released on Sunday 27 February increased to 3300 cubic metres per second on Wednesday, 2 March,[282] so as to have swiftened the drainage of the flood compartment thanks to the extreme weather departing South-East Queensland to be replaced by sunny conditions far more conducive to muscular outflows.

Commensurately, the Brisbane River's height at the City Gauge fell from its Monday 28 February 2022 peak of 3.85 metres and never regained that height – being the 5th highest since 1841 – thanks to the fining up of the weather.

The 'rain bomb' incidentally was not finished; rather than peter out, it continued south to wreak destruction and havoc on Northern NSW, especially to the Northern Rivers city of Lismore, and further south as far as the north-western Sydney Hawksbury and Nepean River systems.

Regardless, the contrast with the 2011 rate of release is striking: 7000 cubic metres per second,[283] more than double, including a re-

---

280 'Wivenhoe Dam did do its job', Letter to the Editor, Neil Brennan of Seqwater, *The Sunday Mail*, 13 March 2022, p. 71.

281 Lydia Lynch, 'Dam overflow stepped up as rains return', *The Australian*, 3 March 2022, p. 5.

282 Ibid.

283 Lydia Lynch, 'Dam overflow stepped up as rains return', *The Australian*, 3 March 2022, p. 5.

lease of 645,000 megalitres on Thursday 11 January 2011 alone[284] rather than the not dissimilar 700,000 megalitres being released over a much more ordered five days between the final weekend of February and Wednesday, 2 March 2022.[285]

Whereas there was no contrast so far as the identical damage and loss wrought to Brisbane River affected suburbs like Goodna and Rosalie as referenced previously was concerned.

Why was the loss and damage the same despite the Wivenhoe Dam releases being so much more gradual and moderated in 2022?

The reason why being, yet again, the unavoidable impact upon Brisbane of 48.461 percent of the Brisbane River's sub-catchment inflows being sourced from sub-catchments downstream and hence wholly independent of Wivenhoe Dam:

- This writer cited one year after 2011 '*the empirical impact upon the Brisbane River ... of water systems downstream ... that Wivenhoe Dam had zero impact at all in holding back ... for the simple reason that the waters added to the Brisbane River from ... the Lockyer Creek* [and]... *the Bremer River, both ... intersect with the Brisbane River downstream from Wivenhoe*'[286];

  and

- The NSW Court of Appeal's noting '... *that slightly less than half of the peak flow at Moggill was attributable to releases from Wivenhoe ... Even if there were zero releases from Wivenhoe, there would have been urban inundation; the peak flows at Moggill without Wivenhoe exceeded 5,000m3/s*'.[287]

Or even non-catchments; the topography of inner-western Brisbane places Rosalie in an unfortunate position: effectively at the bottom of a set of high ridges on which the suburbs of Paddington and

[284] Hedley Thomas & Milanda Rout, 'History forgotten in rush to riverfront luxury', *The Australian*, 18 January 2011, p. 1.

[285] Lynch, op. cit,, p. 5.

[286] *Tennyson Breach*, pp. 115-116.

[287] Appeal, at [653] & [684].

Bardon are located. Meaning that the extreme rainfalls of Sunday 27 February 2022 literally tumbled down steep roadways such as Ellena Street and Fernberg Road into Rosalie's Beck Street, the stormwater drains of which could not convey waters away due to the already overloaded Brisbane River causing the same backing up effect witnessed to houses and businesses adjacent to Enoggera Creek.[288]

Still though the question soon after, ie early March 2022, begged: Should the Wivenhoe Dam releases have happened earlier than 4.00am on Sunday 27 February 2022 and more aggressively?

Such was the tenor of a *Sunday Mail* opinion piece published on 6 March 2022, which the 13 March 2022 published letter from Seqwater's CEO Neil Brennan was responding to, and hence crossed, inadvertently, with an editorial of *The Courier-Mail* published one day earlier expressing markedly different themes and tones to Brennan's:

> *Flood mitigation procedures are simply not good enough. Premier Annastacia Palaszczuk keeps talking about how Seqwater "followed the dam manual, every protocol" during the crisis. That is true. But the manual is broken*' and '*needs to change so that flood mitigation can begin before the Wivenhoe Dam reaches full capacity. Seqwater needs to be able to release massive amounts of water before the Brisbane River is so swollen that any extra water causes greater flooding.*[289]

Exactly what was meant by '*massive amounts*' is not known; no figures were particularised.

Whereas figures were particularised during Seqwater's endeavours over the next week to have drained 700,000 megalitres[290] from Wivenhoe's 2.08m megalitre flood-storage compartment by Wednesday, 2 March 2022. And moreover done so in a fine weather context so as to not have re-flooded the Brisbane River to any height

---

288 Supra Chapter 2.

289 Editorial, 'We'll be paying for floods a long time', *The Courier-Mail*, 12 March 2022, p. 46.

290 Lydia Lynch, 'Dam overflow stepped up as rains return', *The Australian*, 3 March 2022, p. 5.

at the City Gauge exceeding its Monday 28 February 2022 peak of 3.85 metres.

Nor did the editorial acknowledge what the NSW Court of Appeal recognised: the inherent conflict between a flood mitigation strategy on the one hand and a drinking water retention strategy on the other. That the editorial was published a mere 11 days after that same paper quoted Seqwater's Mike Foster making the rather obvious point that '[O]*nce we make a decision to release water from a dam you can't get it back,*'[291] was puzzling to say the least.

Chapter 5 and its sub-chapters essayed the analysis the NSW Court of Appeal performed of the Manual extant as of the January 2011 flood event. Including their Honours' commentary that when taking into account on the ground downstream observed effects, '[T]*he 4,000m3/s rate in the Manual is scarcely precise*'[292] because of the observed evidence at 00:45 on 10 January 2011 '*that in the circumstances which actually prevailed, flow rates at Moggill above 3,500m3/s were damaging*'.[293]

The current version of the manual notably walks these integers back, and incidentally does away with the W1, W2, W3 and W4 strategy levels that also applied in 2011:

- Far from 4000m3 per second being the upper limit of non-damaging effects downstream, 3000-4000m3 per second flows at Moggill are predicted to cause flooding of '*several hundred residential and commercial premises (many with above floor flooding)*'; and
- Instead the new maximum square metreage per second rate capable of flooding around 100 residential and commercial premises (some with above the floor flooding) is only half what the old version provided: 2000 – 3000m3 per second.[294]

[291] Hayden Johnson, 'One dam huge problem', *The Courier-Mail*, 1 March 2022, p. 5.

[292] Appeal, at [627].

[293] Appeal, at [627].

[294] Revision 16 of the Manual, released November 2021, p. 30.

Therefore proving that releasing '*massive*' amounts of water from Wivenhoe Dam, as editorialised, will have many more risks and consequences than forfeiture of drinking water for a thirsty greater Brisbane alone. The Twin Bridges Recreational Area causeway bridge, which links the townships of Fernvale and Wivenhoe Pocket, is the first downstream of Wivenhoe bridge-crossing of the Brisbane River. These four field trip pictures were taken by the author's passenger of the southbound and northbound approaches respectively to the Twin Bridges Recreational Area causeway bridge during early June 2022 and the recreational area generally. The 'recreational' aspect of the area's name is vividly depicted as having been substantially and adversely affected by the forces of the waters, though the intrepid people in the two white four-wheel drives seemed determined to make something of a sunny June Sunday afternoon regardless:

The contention underlying these four pictures is that they depict clearly more than just the Brisbane River causeway bridge as having been inundated; local Fernvale sources interviewed confirmed to this writer that the landscape scarring evident at the heights of both of the Twin Bridges photographs' backgrounds represented the peak of the March 2022 Brisbane River flows. It is a matter of record that the maximum flow rate of outflows in the first week of March 2022 were 3300m3 per second.[295] Whereas the natural flow rate at that same area including flows from the upstream confluence of Lockyer Creek during the phase Wivenhoe Dam's gates remained completely shut save for the one solitary hour on 25 February 2022 did not exceed 1600m3 per second.[296] Proving that the deleterious environmental effects depicted by the four previous photographs were not due to natural rhythms of the area, but rather the deliberate Wivenhoe Dam outflows which reached a maximum of 3300m3 per second.

It is therefore submitted that '*massive*' flows out of Wivenhoe Dam will risk not just drinking water being forever forfeited, but

---

[295] Lydia Lynch, 'Dam overflow stepped up as rains return', *The Australian*, 3 March 2022, p. 5.

[296] Rob Ayre, interview with the writer.

much in the way of downstream environmental damage also. Not inconsistently the Manual expressly does not allow for pre-emptive releases based on forecasts because forecasts do not provide sufficient assurance for ensuring that Wivenhoe's water supply component will be completely full at the end of a flood event.[297]

This aspect of the Manual is, to be frank, not '*broken*', nor in any other way in '*need* … [of] *change*'.

To the contrary, that aspect of the Manual is eminently sensible and logical.

The usage of the word '*paying*' in *The Courier-Mail*'s editorial of 12 March 2022 was a harbinger of its underlying theme: an insurance bill, estimated at the time, of in excess of $1 billion. The paper had, entirely appropriately, devoted reams and reams of newsprint and photos to the human cost and toll stories of devastation caused to myriad residential houses and small businesses during and then in the long, hard, hot, humid and dirty days of cleaning up the mud and filth left behind as the flood waters eventually receded. That a degree of emotion necessarily novated its way into an editorial penned almost two weeks afterwards is explicable and understandable. However objectivity must remain the touchstone for a newspaper of record being the only hard-copy publication for the city of Brisbane. Especially when considering the cited insurance bill of $1 billion context for that editorial, a substantial component of which will necessarily be devoted to the devastation wrought to Brisbane's northern suburbs as a result of what that same paper correctly editorialised 13 days earlier were '*historic flows in pretty well every river and creek across the South-East*', some of which were wholly unconnected to the Brisbane River, let alone Wivenhoe Dam.

And Wivenhoe Dam's Manual.

Just as the class action lawsuit against Seqwater was ultimately not made out, *The Courier-Mail*'s informal 'judgment' against Seqwater's role in 2022 similarly fails.

---

297 Revision 16 of the Manual, released November 2021, p. 58.

7

Weather Forecasting

Another omission from *The Courier-Mail*'s informal 'judgment' against Seqwater's role in 2022 was the role of weather forecasting.

An extremely serious role at that.

In the NSW Court of Appeal judgment, after a meticulous analysis of the 80 or so pages of the Manual extant as of January 2011, their Honours concluded:

> *Essentially, the procedure described in the Manual involved the engineers predicting as best they could what the* ***likely maximum amount*** *of water was going to be. Then …they were to regulate outflows by reference to the primary consideration as identified by the Manual. That is to say, the Manual answered the basic question a flood engineer must ask during a flood event:*
>
> *"Given the* ***likely maximum amount*** *of water in this dam, should I be focussing on protecting the dam structure, or should I focus on protecting urban infrastructure, or may I merely focus on protecting downstream bridges."*[298]

The concept of '*likely maximum amount*' necessarily introduces weather forecasting:

> *… it is tolerably clear that the estimation of the likely level of water in Wivenhoe Dam is to be informed by rainfall forecasts … many other references to rainfall forecasts and predictions* [exist] *in the Manual … If predicted dam levels were to exceed 74m, then the primary objective was to preserve the dam. If predicted dam levels exceeded 68.5m, then the primary objective*

[298] Appeal, at [233] – bold font added.

was to prevent urban inundation. If the predicted dam levels were lower, then subordinate flood mitigation objectives (such as preserving the downstream bridges) came to the forefront.[299]

Of course weather forecasting is an inherently difficult endeavour, especially for '... *dam operators* [who] *do not have the gift of foresight*'[300] or for downstream administrators forced to deal with the aftermath:

Where the weather is concerned, forecasts necessarily are tentative and this 2022 event was rapidly evolving. One has to be careful about being judgmental about what [the Brisbane City] *Council should or should not have done. It is easy to use hindsight to stake the moral high ground, but one has to be realistic and there are certain things that simply could not be accomplished bearing in mind the rapidity of the event.*[301]

At the height of the 2022 event, Seqwater's Mike Foster was quoted:

Sometimes the forecasts aren't right, sometimes the forecasts can be really, really wrong ... Once we make a decision to release water from a dam you can't get it back ... we could find ourselves in a situation where we release large volumes of water where we've flooded Brisbane unnecessarily based on a forecast.[302]

In other words, a dam operator could release large volumes of water pre-emptively based on high rainfall and therefore inflows forecasts, in controlled ways that do not cause downstream flooding, yet create an entirely new problem in the process.

Because drinking water once released is drinking water forever forfeited:

... should ... forecast rainfall be less than predicted, or indeed not occur at all, then any pre-released water could not be re-

---

[299] Appeal, at [205] and [227].

[300] QFCI Final Report, p. 438.

[301] The Hon Paul de Jersey, *Brisbane City Council 2022 Flood Review*, 9 May 2022, p. 9.

[302] Hayden Johnson, 'One dam huge problem', *The Courier-Mail*, 1 March 2022, p. 5.

> *covered ... this would instead unnecessarily expose the city* [of Brisbane] *to the risk and consequences of drought, including water restrictions and operation of desalination plants.*[303]

Given that only 15.08% of drinking water supply for Brisbane was remaining for a thirsty city during 2007, ensuring sufficiency of supply, and therefore not wasting, dam water held for drinking water purposes is essential. I didn't devote an entire chapter of my previous work to '*The Politics of Drought*' for nothing; enough content for an entire chapter existed, including claims that then Prime Ministerial aspirant – for the first time – Kevin Rudd was the '*architect of Brisbane's drought*'[304] due to the cancellation, essayed in Chapter 4, of the Wolffdene Dam proposal.

Consider then if the rain bomb didn't in fact fall as it did in February 2022, but '*massive*' amounts of water were released from Wivenhoe Dam into the Brisbane River in anticipation of it? Consider also that no decent rain in the four year period to February 2026 thereafter falls in the Upper Brisbane and Stanley catchments which directly feed Wivenhoe, causing Wivenhoe's drinking water capacity to recede to 15.08% again. How would an editorial writer, in March 2026, have characterised such a calamity of errors? Not sympathetically that's for sure.

Of course politics would inevitably intrude into the 2022 flood event just as it did into the 2000-2007 drought; under the usual pressure that comes to be visited upon politicians after natural disasters, then Queensland Premier Annastacia Palaszczuk '*forcefully defended authorities against community anger that they weren't warned soon enough of the unfolding flooding disaster across South-East Queensland ... "It has been fast and it has been furious and it had a big impact ... What happened here was that everyone expected the conditions to ease, but they didn't and this rain bomb stayed over the entire South-East and it had a big impact on the catchments and*

---

303 Dr Luke Toombes and Rob Ayre, 'Resetting Expectations of Brisbane's Flood Protection', p. 11.

304 *Tennyson Breach*, p. 74.

the streams...that is unpredictable. No one, not even the bureau [of Meteorology] *saw that coming*".[305]

The Premier was correct. Forecasts of rain in the lead up to the 2022 event were, not unlike the forecasts cited by the NSW Court of Appeal at the beginning of January 2011, very anodyne: '*Brisbane Seven-Day Forecast*' published on Tuesday 22 February 2022 only cited '*heavy falls*' as possible on Thursday 24 February; the big three deluge days of Friday – Sunday 25 – 27 February 2022 were forecast to be 26, 28 and 29 degrees centigrade with '*Showers*', '*Shower or two*' and '*Shower or two*' the respective forecasts.[306]

'*Shower or two*' and '*rain bomb*' being rather mutually exclusive concepts.

Note too that the temperatures actually recorded on Friday – Sunday 25-27 February 2022, due to the heaviness of the cloud coverage blocking out the heat of the sun, were commensurately lower than forecast: 24.2, 23.9 and 23.6. Almost cold-snap style temperatures for Brisbane during the month of February. Even on one winter day during 2022, Brisbane exceeded the final of those three temperatures![307]

Admittedly the weather forecasts rapidly became far more serious and closer to accurate towards Brisbane's unforgettable final weekend of February 2022, the epithet '*slam*' being especially prophetic: '*... a deadly storm system could slam South-East Queensland Thursday and Friday amid warning to expect more than 300mm of rain*' was reported two days later, Thursday 24 February 2022.[308]

Yet even then the more than 300mm of rain expectation was out by more than double, given that 793mm actually fell.

The forecasts of 22 and 24 February 2022 respectively are there-

---

[305] Jessica Marszalek, 'I can't control Mother Nature', *The Courier-Mail*, 1 March 2022, p. 5.

[306] *The Courier-Mail*, 22 February 2022, p. 39.

[307] 23.8 degrees centigrade being recorded on 22 June 2022: Bureau of Meteorology Brisbane, June 2022, Daily Weather Observations.

[308] Felicity Ripper, Tegan Annett, Kristen Camp, 'Deluge getting stronger', *The Courier-Mail*, 24 February 2022, p. 4.

fore apposite examples of the ways in which the NSW Court of Appeal analysed the rather serene, relatively speaking, rainfall forecasts issued on 2 January 2011 compared to the '*biblical proportions*' of rain which actually fell some days later. The thesis being that the longer the range of forecast, the greater the probability of the forecast not being fulfilled.

Which, of course, was the result in 2022; the actual seven day Brisbane rainfall was in fact an all time record, with parts of the Somerset and Wivenhoe Dam catchments receiving a year's rainfall in four days.[309] Little wonder the '*rain bomb*' analogy was conceived. Again some may decry the crassness of the term, however that was far more accurate a descriptor than '*shower or two*'.

Of course weather forecasts can and do turn out to be just as equally erroneous or much milder in reverse; disappointment is palpable when, during phases of drought, predictions of desperately needed rains do not eventuate.

Take cyclones for instance. Cyclones form and are energised by warm sea waters, the Coral Sea being notorious as a breeding ground for many systems, whether small, medium, large or monster: the Category 5 behemoth Tropical Cyclone Yasi of early February 2011 being one requiring no introduction. Cyclones begin as low pressure systems – areas where atmospheric pressure is lower than surrounding areas, causing both air and water vapour to rush into the area; the stronger the low the inevitable it is they form cyclones.[310] Once cyclones cross the coast they lose their energy and become less destructive, though the remnants of the systems form what are known as rain depressions. The 1893 and 1974 Brisbane River floods were ultimately spawned by the rain depressions that Cyclones Althea and Wanda became after they made landfall.

Therefore for a floodplain city like Brisbane, fed by a river as long as the Brisbane River is with a total catchment area of some

---

309 Hayden Johnson, 'One dam huge problem', *The Courier-Mail*, 1 March 2022, p. 5.
310 Stuart Layt, 'Cloudy with a chance of hyperbole', *The Saturday Age*, 9 April 2022, p. 30.

13,570km2[311] combined with its many tributaries, any tropical cyclone which crosses the Queensland coast south of the Tropic of Capricorn could conceivably dump vast amounts of rain and create yet another flood event.

Being Brisbane born and raised, I have not lived through a cyclone. The best I can offer up recollection wise in my lifetime so far as potential impacts were concerned are Cyclones Nancy in 1990, Hamish in 2009 and Oma in 2019. Nancy and Oma caused Brisbane to be placed on actual cyclone watch: *ABC News 24*'s strap line lit up in red coloured warning font '*Brisbane on cyclone watch for the first time in 29 years*' during February 2019 when Cyclone Oma began approaching South-East Queensland. The Queensland Fire and Emergency Service went one better warnings wise by casting the system as being the opposite of the usually avuncular way that grandmothers are perceived:

Queensland Fire and Emergency Services - QFES
February 20, 2019 ·

Did you know Oma means grandmother in German and Dutch? Well, this old lady isn't visiting to bake cookies!

Cyclone Oma is making herself known with strong gales and heavy rainfall along the south-east coast.

Make sure you're prepared by keeping up-to-date with the latest warnings - particularly if you're planning a trip to the beach. As we near the weekend, expect big waves, windy conditions and lots of rain.

Don't ignore Oma and consider staying indoors this weekend (with some knitting, of course).

---

311 *First instance*, Ch 2, at [39].

In the result though, Oma, nor Nancy and Hamish either, crossed the coast. Cyclone warnings are forecasts nonetheless. Forecasts can and do turn out differently. Considering the ways in which non-cyclonic weather systems wrought the damage to Brisbane that the 2011 and 2022 rain events did, it is undeniable that Cyclones Nancy, Hamish and Oma's collective die outs at sea, albeit contrary to certain track map models which forecast actual coastal crossings, were for the best.

Just as equally for the best was that the engineers tasked with operation of Wivenhoe Dam did not pre-emptively release water, let alone the '*massive*' quantities advocated by *The Courier-Mail* editorial on 12 March 2022, on the basis of track map models which forecast actual coastal crossings of those three cyclones and commensurately outsize rainfall forecasts upon making landfall. Including one rogue model which predicted a coastal crossing by Cyclone Oma with a consequential seven-day depth of rainfall of over 1.5 metres.[312] The actual result during the week the subject of that model forecast was far more prosaic: the Brisbane rainfall gauge recorded 0.4mm, while the Wivenhoe Dam gauge received nothing.[313] A true allegory for what a dry and dusty year 2019 became. Therefore had any pre-emptive release from Wivenhoe taken place, Brisbane's existential pendency as a capital city could have been imperilled due to the sheer paucity of rainfalls in Brisbane and over the Wivenhoe Dam catchment during Oma's 2019 year of brief existence.

A similar set of debates ensued during August, September and October 2022 as fears of onset of another deluge or set of deluges in the upcoming summer of 2022/23 emerged as a result of conducive to high rainfall quantities weather system forecasts. Though Brisbane's 2022 winter followed a conventional script with only unremarkable rainfalls recorded between June and mid-August 2022,[314] a newspaper headline of '*Prepare for six months of rain*' and sub-

---

[312] Dr Luke Toombes and Rob Ayre, 'Resetting Expectations of Brisbane's Flood Protection,' p. 11.

[313] Ibid.

[314] Compared to May 2022 – see Chapter 9, 'May Mayhem' below.

headline of '*We're the 'meat in sandwich' of La Nina flood alert*' caused residents to be abruptly disabused of any hope of that trend lasting into the months of the summer of 2022/23. A co-incidence of a Pacific Ocean in La Nina phase and negative Indian Ocean Dipole meant that '*both major oceans on either side of Australia are in a wet phase, a double negative, which will enhance rainfall*'. Annastacia Palaszczuk was moved to implore Queenslanders to prepare:

> *I don't want Queenslanders to be alarmed, but what I do want to see is people to be prepared... We will talk to all of the mayors just to make sure we're very well prepared.*[315]

Sky News chief meteorologist Tom Saunders added that since 1900 there had only been two double-negatives in the twin oceans: 1974, 2010 and 2021.[316] By this juncture in this work, readers ought already know what occurred either during or immediately after those three particular years.

Hence it was not surprising that

> [T]*he Bureau* [of Meteorology] *report*[ed] *a negative Indian Ocean Dipole event is underway ... The Bureau's El Niño–Southern Oscillation (ENSO) Outlook moved from La Niña WATCH to La Niña ALERT ... As such, the Bureau advises there is a very high chance of wet conditions over eastern Australia for the next three months. If a La Niña event is established, wet conditions will persist into the summer of 2022/2023 ... In light of the Bureau's recent outlook for above average rainfall, wet soils, high rivers and full dams, an elevated flood risk remains for eastern Australia.*[317]

Brisbane's Lord Mayor began the dam policy debate by asking, on LinkedIn '[I]*s it time for State Government dam operators to pro-*

---

[315] Natasha Innes, 'Prepare for six months of rain', *The Courier-Mail*, 16 August 2022, p. 6.

[316] Ibid.

[317] Office of the Inspector-General Emergency Management, 'South-East Queensland Rainfall and Flooding February to March 2022 Review', 31 August 2022, p. 39 of 155.

actively lower dam water levels before another potential season of severe La Nina conditions?'[318] One month later Schrinner's question was partially answered by both the Premier and Water Minister for Queensland, based on '*briefings and advice from Seqwater, the Department of Environment and Science, and the Department of Regional Development Manufacturing and Water*', this way:

> *In preparation for this summer's forecast La Nina, Wivenhoe Dam's temporary full supply level will be lowered... to 80 per cent... to ensure Queenslanders and communities remain safe... minor gated releases will occur from Wivenhoe Dam ... Across the next two weeks, Seqwater will drain around 116,000 megalitres of water.*[319]

Wivenhoe Dam's drinking water storage compartment, pre-announcement, was at 90.5 percent capacity.[320] This is what the practical result of reducing the storage from there to 80 percent looks like {the planned nature of the release drew sightseers out to the Dam's observation deck to watch and photograph the event}.[321]

---

[318] Adrian Schrinner, Brisbane Lord Mayor, September 2022, via LinkedIn.

[319] Joint Statement Premier and Minister for the Olympics, The Honourable Annastacia Palaszczuk and Minister for Regional Development and Manufacturing and Minister for Water, The Honourable Glenn Butcher, 'Wivenhoe Dam readied for La Nina', Thursday, 13 October 2022, at 01:59 pm.

[320] Lydia Lynch, 'Dam balancing act of flood and drought', *The Australian*, 3 October 2022, p. 2.

[321] Footage credit: Andrew Clarkson.

A close noting of Clarkson's footage will depict that only one of Wivenhoe's five radial gates opened – note the daylight visible on top of the two left and two right gates not being repeated for the median of the five.

Presumably in a bid to prevent complete wastage of those 116,000 megalitres by utilising the specifically released waters for domestic purposes, all South-East Queensland households connected to the SEQ water grid would have their water bills for the billing period including October 2022 rebated by $55: '*By applying a $55 discount, we want to encourage families to get out and make the most of this opportunity. Now is the time to wash your driveway, gurney your house and driveway and flush out your gutters*',[322] Minister Butcher recommended.

Though the combined Palaszczuk and Butcher announcement conflicts with the Manual's prohibition of pre-emptive releases from Wivenhoe's drinking water compartment, Seqwater '*is controlled by the Queensland Government which is its ultimate parent*'.[323] Meaning that the political heat or kudos, depending on whether or not the lost 116,000 megalitres are replaced over the summer of 2022/23, will be that of the State, not Seqwater or its board, management or engineers.

In conclusion to this chapter, the conflict identified by the NSW Court of Appeal between Wivenhoe Dam attempting to wear simultaneously a 'flood mitigation' hat and a 'maximising of drinking water supply' hat, is therefore very real.

And when the management of that conflict is to be informed so substantially by both long-term and short-term rainfall forecasts, with the benefit of 2011 and 2022 hindsight, who would want to become a dam operations engineer, or a responsible state government minister for dams, today?

---

[322] Joint Statement Premier and Minister for the Olympics, The Honourable Annastacia Palaszczuk and Minister for Regional Development and Manufacturing and Minister for Water, The Honourable Glenn Butcher, 'Wivenhoe Dam readied for La Nina', Thursday, 13 October 2022, at 01:59 pm.

[323] Seqwater Annual Report 2021-22 ISSN: 1837-4549, p. 32.

8

Planning, Planning Policies and Circularities

The $1 billion insurance bill cited in *The Courier-Mail*'s 12 March 2022 editorial was hardly likely to be warmly received by the insurers themselves.

And that was just the beginning of the prognostications; by 25 May 2022 107,562 insurance claims for the State of Queensland overall had been lodged, with the southern Queensland component value total reaching $1.82 billion.[324] Andrew Hall, CEO of the Insurance Council, when interviewed on 13 June 2022 described the 2022 flood event in both South-East Queensland and Northern New South Wales '*as the fourth most costly natural disaster in Australian history, with a $4.3 billion price tag of losses*'.[325]

As it transpired, the longer one waited the greater the South-East Queensland and Northern New South Wales event's ranking in the score of most costly Australian natural disasters seemed to become:

- By 28 June 2022 that fourth most costly estimate had firmed to third (at $4.8 billion, based on 225,000 claims lodged from both South-East Queensland and Northern New South Wales, valuing the February/March 2022 event behind only Cyclone Tracy in 1974 and the 1999 Sydney hailstorm as the worst ever insurance wise[326]); then
- By 8 September 2022 from third to second, at a total cost of $5.28 billion, $1.38 billion of which was attributed to Brisbane's loss and damage alone[327]; and, by year's end,

[324] Felicity Ripper, '*Deluge forces big cash splash*', *The Courier-Mail*, 22 May 2022, p. 5.

[325] Michael Atkin, ABC *7:30* telecast, 13 June 2022.

[326] Ciara Jones, *ABC News* Qld telecast, 28 June 2022.

[327] Rachael Rosel, 'Our most expensive ever flood', *The Courier-Mail*, 8 September 2022.

- $5.65 billion, exceeding the $5.57 billion Sydney hailstorm, as a result of a total of more than 237,000 claims having been lodged by December 2022.[328]

Suncorp is Queensland's largest company by stock market capitalisation. It also extensively sponsors sports and events, making it an icon of this state, especially one that has always borne a stigma of being a merely 'branch office' town compared to Sydney, Melbourne and Perth.

Not surprisingly then, Suncorp's CEO Steve Johnson was approached for comment a matter of days after the flood peak, '*blasting*' a century of planning laws which put '*people in harm's way*' by allowing construction activity in inadequate locations.[329] Johnson added that Suncorp had lost money on home cover policies for the past five years,[330] which is quite a noteworthy statement given that the drought blighted conditions that endured during 2017-2019 would be expected to have reduced claims stemming from water damage caused by floods and/or high rainfalls. Speaking of the insurance market generally, the ABC's Michael Atkin reported that '[I] *nsurers have recorded a staggering underwriting loss on home insurance of $420 million in the past two years*'.[331] The Insurance Council cited inflation, increased building material prices and rises in global reinsurance premiums as contributing to a nationwide '*upward pressure on insurance premiums*'.[332] There is therefore only one way private sector insurance companies can recoup those losses: higher premiums. Usually, and unfortunately, from cohorts of persons least able to afford them. Assuming they even want to take out insurance in the first place that is; Melbourne Law School survey and analysis

---

[328] Michael Read, 'East Coast flood nation's most expensive natural disaster', *The Australian Financial Review*, 1 December 2022, p. 3/Ben Cubby, 'Evacuations as record flood threatens Menindee', *The Sydney Morning Herald*, 31 December 2022, p. 7.

[329] Madura McCormack, 'Insurer blasts bad planning laws', *The Courier-Mail*, 4 March 2022, p. 10.

[330] Jamie Walker, 'A Blueprint for Resilience', *The Weekend Australian*, 5-6 March 2022, p. 31.

[331] Michael Atkin, ABC *7:30* telecast, 13 June 2022.

[332] Ibid.

has theorised that post-natural disaster insurance claims processes are, for some claimants, more stressful than the extreme weather events which actually created the damage.[333] That being the case, less, not more, premium revenue will flow to insurers if former insureds are in fact so disaffected by the claims process that they elect not to renew.

Of course dwelling locations 'inadequate' in a flooding sense can nonetheless be adequate in many other senses, including price. Goodna is such an area – close to shops, medical centres, schools, roads and the railway line connecting Ipswich to Brisbane. However it is inherently, almost hopelessly, flood prone, having been devasted in both 2011 and 2022, despite the differing dynamics of those two flood events analysed in chapter 6. In a bitter illustration of the famous statement that the Sydney Harbour Bridge requires so much painting that once the end-to-end paint job is finished, it's time to start all over again, a Goodna resident '*had only just finished fixing his home from the 2011 flood when it was completely submerged last week…the swollen river almost completely swallowed his house*'.[334]

Goodna's Achilles heel is its location at the confluence of the Brisbane River and Woogaroo Creek, the implications of which having been explained in Sub-Chapter 5.11. Not surprisingly therefore that for the entire Ipswich City Council jurisdictional area, 224 out of a total of 688 2022 flood event property damage counts were suffered in Goodna alone – 32.56% of Ipswich's overall damage to just 1 of its 22 suburbs.[335]

Rocklea, an inner-southern suburb, is closer still to the Brisbane CBD, and, like Goodna, is also serviced by a direct railway line into the inner city, making it, on paper, an even more attractive residential proposition. In addition, Rocklea is in close proximity to some

---

333 Jill Rowbotham 'Insurers "worse than the disaster"', *The Australian*, 23 August 2022, p. 5.

334 Charlie Peel, 'Homeowners seek buybacks for in-danger properties', *The Australian*, 10 March 2022, p. 5.

335 Qld Fire & Emergency Services 2022 Damage Assessment Dashboard for the Ipswich Local Government Area, extracted p. 14 of the Ipswich City Council 2022 flood review.

of Brisbane's major public hospitals such as the Princess Alexandra, Mater and the Queensland Childrens' Hospital. For one of the 73 year old Cambridge Street Rocklea residents Atkin interviewed, the fear of moving further away from medical care seemed a greater risk than living in the same house so badly damaged in both 2011 and 2022, even when forced to convert their garage to a makeshift living room and kitchen: '*As we're getting older the hospitals and everything are closer*'.[336] The inferential trade-off for that interviewee being that, to sell and purchase something genuinely flood proof, would involve moving much further away from hospital care than the actually quite conveniently located, in all senses bar flooding, Rocklea.

Other Rocklea residents took proactive steps post-2011 such as rebuilding with tiles and cement sheeting which enabled one to '*rid*[e] *out* [last] *Sunday's inundation*'[337]. Whereas another resident merely '*considered*' raising his house off a ground level slab when interviewed during January 2021 for the 10 year anniversary of the 2011 event, proclaiming at the time '[I]*f it happened again I'd swear about it and then do it all again*'.[338]

Undoubtedly then much in the way of expletives permeated Rocklea during the final weekend of February 2022. Not that the prone to cussing resident Jamie Walker interviewed immediately afterward was motivated to, once and for all, leave, vowing, instead, to rebuild.[339]

Residents had their chance to have departed however; following the 2011 floods the Brisbane City Council invested approx. $58.4 million in the purchase of 112 houses in Rocklea between then and 2016[340] when the program ceased with no prospect of program revival absent State and Commonwealth Government support.[341]

---

336 Michael Atkin, ABC *7:30* telecast, 13 June 2022.

337 Jamie Walker, 'A Blueprint for Resilience', *The Weekend Australian*, 5-6 March 2022, p. 31.

338 Ibid.

339 Ibid.

340 Ibid.

341 The Hon Paul de Jersey '*Brisbane City Council 2022 Flood Review*', 9 May 2022, p. 58.

The Brisbane City Council appeared to change tack after the buy-back scheme ended, introducing during 2018 its '*Flood Resilient Homes Program*', designed to '*retrofit*' existing flood-prone dwellings to increase their flood resilience.[342] A voluntary program, 286 properties were assessed, works completed in 144 homes (including one house raising) at a cost to ratepayers, so far, of $9,878,860.[343] Following the 2022 flood event, 100 of that total of 144 homeowners were surveyed. 75 of whom reported that floodwater impacted the area of the resilience work, and 62 of those 75 (83% of the sample size) vindicated the resilience works as having been successful.[344]

Three days after the delivery of the de Jersey report, a $741 million joint Queensland and Federal Government package was announced by the Queensland Government enabling home owners to apply for grants to rebuild and raise homes or buy back high risk properties: '*Those who choose to stay can gain access to grants that replace floor coverings with more flood-resilient finishes like tiles or polished concrete. Power outlets can be raised. It is entirely up to the homeowner*' commented Ms Palaszczuk.[345]

The choices being[346]:

- Resilient household rebuild program – repairs or retrofit flood resilient design and materials (up to $50,000);
- Home-raising program – funding to elevate liveable rooms above the flood level or relocate within blocks (up to $50,000); or
- Voluntary home buy-back program where a determination is made that repair, retrofitting or raising is not suitable. Applications under this program were to be considered on a

342 Ibid., p. 56.

343 Ibid.

344 Ibid., p. 57.

345 Taylah Fellows Jessica Marszalek, 'Flood money up for grabs', *The Courier-Mail*, 13 May 2022, p. 11.

346 Tennyson Ward Office, 'Resilient Homes Fund', emailout 12 May 2022.

> case-by-case basis taking into consideration factors including frequency of flooding, severity of flooding, risks to structural safety and human life and broader social impacts.

By 13 June 2022, 2600 people in total had applied[347]. This number had risen to 5400 by the end of November.[348]

Then Deputy Premier and Planning Minister, and now Queensland Premier, Steven Miles, estimated '*around 500 properties*' where buy-backs were '*the most appropriate outcome*' across South-East Queensland.[349] Seven months later 180 governmental offers had been made, with 131 agreements to sell having been reached, causing Miles to repeat his at least 500 prediction.[350] Given the 112 buybacks actually made by the Brisbane City Council within Rocklea between 2011 and 2016 and the further 131 agreed to during the latter half of 2022, Miles' estimate of 500 appears appropriate

By way of example, this is what Boobook Street in Rocklea looked like during May 2022 following its own set of buybacks; the fencing of nothing being quite a stark emblem:

[347] Michael Atkin, ABC *7:30* telecast, 13 June 2022.

[348] Samantha Healy, 'Flood buy back fund inundated', *The Courier-Mail*, 26 November 2022, p. 24.

[349] Taylah Fellows, Jessica Marszalek, 'Flood money up for grabs', *The Courier-Mail*, 13 May 2022, p. 11.

[350] Jack McKay, 'It's buy buy to 131 flooded Qld homes', *The Courier-Mail*, 5 January 2023, p. 15.

Other bought back houses on Boobook Street became, by effluxion of time, unofficial community parkland. Unofficial because no name designation of parkland was made, nor the usual infrastructure such as playground equipment, benches and barbeques being installed:

That could be because three of the 112 buy-backs which were converted to a children's playground came to be '*coated in putrid mud*'[351] during the February 2022 event. Council reluctance to invest any ratepayer funds at all in locations bought back for reason of their guaranteed liability to be flooded is explicable.

I visited Boobook Street in Rocklea on Saturday 14 May 2022, a day that fell during another high rainfall event for Brisbane.[352] The march of floodwaters at the Stable Swamp Creek end of that street is palpable despite barely more than 100mm having fallen in the 13 days leading up to the 14th day of May:

---

351 Jamie Walker, 'A Blueprint for Resilience', *The Weekend Australian*, 5-6 March 2022, p. 31.

352 See Chapter 9, '*May Mayhem*', below.

The intense saturation the surrounding catchment sustained 2 ½ months earlier obviously was going to be retained within such a short period of time afterwards, meaning less ability for Rocklea's Stable Swamp Creek to convey away waters rained down:

However the May 2022 flooding in and around Brisbane was very minor compared to February. That such a comparatively small amount of rainfall in an overall sense could hem in those home-owners at the end of Boobook Street indicates the sheer lack of feasibility of certain properties ever remaining capable of housing people.

A similar private house to parkland trend will emerge in Goodna, the other of South-East Queensland's two most haplessly flood affected residential suburbs. The owner of a Cox Street Goodna property submerged to its ceiling in 2011 accepted a State Government buyback offer, with Ipswich City Council Mayor Teresa Harding saying its house would be razed and added to the footprint of existing parkland.[353]

---

[353] Madura McCormack, 'Dozens to miss out on $350m buyback', *The Courier-Mail*, 17 October 2022, p. 10.

May 2022 also saw the publication of a Climate Council report with the foreboding title '*Uninsurable Nation*', listing Australia's 10 most climate-vulnerable places. Four of which were the Brisbane, Gold Coast, Lockyer Creek and Scenic Rim local government areas. The Climate Council report's claim was that Queensland is on '*a high risk trajectory of becoming the most collectively uninsurable state*' in Australia by 2030,[354] with a total of 520,994 properties nation-wide being at risk of uninsurability within eight years.

200,000 of that 520,994 number being properties in Queensland, meaning '*Australia's growing home insurance affordability crisis is set to hit Queensland harder than any other state*'. Ipswich Mayor Teresa Harding cited Goodna constituents of hers demonstrating their most recent home insurance bill having reached $11,400.[355] Though at least they could, theoretically, obtain insurance. Not even the public interest nature of insuring a medical centre could convince any insurer to cover Dr Bee Kho's Rocklea general practice as a result of the devastation his area suffered during 2011, meaning he and he alone was required to incur $200,000 of debt after the muddied waters of February 2022 intruded Dr Kho's surgery and destroyed his computers, fridges, examination beds and equipment.[356]

Proving if a medical centre can not obtain insurance, what hope do society's battlers have? (See photos on following page).[357]

There is genuine evidentiary basis for the Climate Council's prediction of increasing climate vulnerability; that the Brisbane River's fourth and fifth largest flood events were separated by only 11 years depicts an obvious contrast when considering, before 2022, that the respective time gaps linking the first, second, third and fourth largest reliably recorded flood events were 52, 81 and 37 years.[358]

These long periods of time between previous floods compared to

---

[354] Sally Gyte, *Seven News* Brisbane telecast, 3 May 2022.

[355] David Mills, 'High-risk Queensland homes will become 'uninsurable' by 2030', *The Courier-Mail*, 3 May 2022, p. 9.

[356] Heather Saxena, 'Flood chaos: "I'm not giving up" ', *Australian Doctor*, 5 May 2022, p. 4.

[357] Photo credit: Dr Bee Kho.

[358] 1841, 1893, 1974 and 2011.

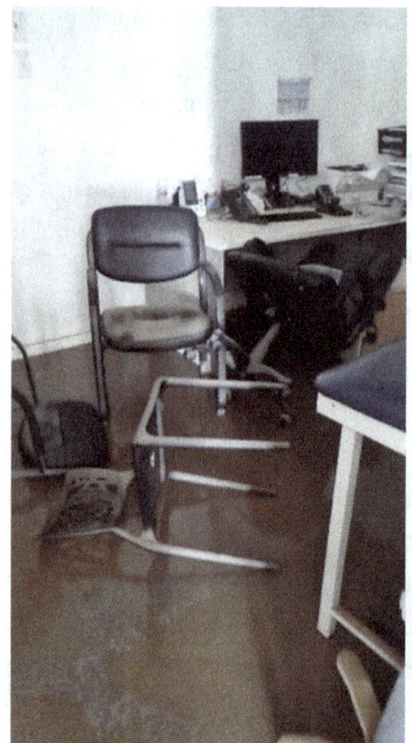

BULK BILLING
07 3493 0880
www.brisbanemarketdoctors.com.au
SKIN CANCER CLINIC
medicare Bulk Billing

2011 and 2022, especially the bisecting of the 37 years between 1974 and 2011 by Wivenhoe Dam's commissioning, meant that '*decades*' were left '*between flood events to recoup losses through insurance taxes and rates, and discouraging the introduction of non-structural flood mitigation. Diminishing flood memories allow complacency, buttressed by the delusion that humans have controlled nature by taming the* [Brisbane] *river*'.[359]

This declaration by Dr Cook as regards an increasingly blasé attitude to planning after previous Brisbane River flood events might, just might, become serious now that the time gap between Brisbane's fourth and fifth largest recorded flood events reduced to only 11 years.

Meaning that the Brisbane River could again reach the 3.85m recorded at the City Gauge on 28 February 2022 as early as the summer of 2022-2023; flood events happen randomly, not according to some pre-set uniform time scale:

> *The Wivenhoe and Somerset dam catchment received a year's rainfall in four days but* [Griffith University Hydrology] *Professor* [Bofu] *Yu urged against thinking it couldn't occur. "Floods tend to be uncorrelated…that means if you've had a flood, it doesn't mean you're not going to have another one".*[360]

Indeed, the Brisbane River, not even 2 ½ months later, exceeded by 11cms its minor flood level of 1.70 metres at the City Gauge as a late and unseasonal weather event added insult to injury to a city that had only just finished mopping up.[361]

A trend is notably emerging:

> [O]*bservations show that there has been an increase in the intensity of heavy rainfall events in Australia. The intensity of short-duration (hourly) extreme rainfall events has increased by around 10 per cent or more in some regions in recent de-*

[359] *River*, p. 196.

[360] Hayden Johnson, 'One dam huge problem', *The Courier-Mail*, 1 March 2022, p. 5.

[361] 1.81 metres at 22:49 on 17 May 2022.

cades, with larger increases typically observed in the north of the country.... As the climate warms, heavy rainfall events are expected to continue to become more intense. A warmer atmosphere can hold more water vapour than a cooler atmosphere, and this relationship alone can increase moisture in the atmosphere by 7 per cent per 1 degree Celsius of global warming... Increased atmospheric moisture can also provide more energy for some processes that generate extreme rainfall events, which further increases the likelihood of heavy rainfall'[362]

Jamie Walker presciently concluded his interview article with Rocklea flood victims by theorising '[W]*hat's inescapable is that severe floods are happening more frequently than ever before and doing unprecedented harm in the process. One way or another, something has to change*'.[363] Presciently because it was written three months before the Bureau of Meteorology published its observational analysis extracted in the preceding paragraph. That Walker's submission that '*something has to change*' was published in the less than emphatic to the concept of climate change induced economic and electricity generation reforms Weekend Australian is noteworthy, as was this opinion piece following the release of the Climate Council report:

It doesn't matter whether you believe in climate change or not because insurance companies do believe in it and are adjusting their premiums accordingly.[364]

Sensing the mood, that paper's environment editor then ventured, following the outcome of the 2022 Federal Election,

... memories of the 2019 bushfires and recent severe flooding on the east coast had an impact on how some people voted ...

---

[362] Bureau of Meteorology, 'Special Climate Statement 76 – Extreme rainfall and flooding in south-eastern Queensland and eastern New South Wales', 25 May 2022, p. 18.

[363] Jamie Walker, 'A Blueprint for Resilience', *The Weekend Australian*, 5-6 March 2022, p. 31.

[364] Katrina Grace Kelly, 'Uninsurable Homes to create a climate of fury', *The Weekend Australian*, 7-8 May 2022, p. 36.

A demand for action on climate change can help to explain the strong showing of the Greens and teal independents ...[365]

Very strong in fact for the Greens; their holding of one solitary House of Representatives seat since 2010[366] quadrupled in 2022 when that party shocked – and glaringly embarrassed – both the ALP and the LNP after claiming the inner-Brisbane federal seats of Ryan, Griffith and Brisbane.

While on the topic of climate change, supplanting coal and gas fired power generation with non-carbon emitting renewables such as wind and solar is an inexorable, albeit weather dependent trend; the effects of the rain bomb firstly in Queensland and then NSW caused wholesale electricity spot prices to have surged in the first three months of 2022 to average $90 per megawatt hour across the power grid compared to $36 per megawatt hour during the first quarter of 2021, according to data held by Origin Energy.[367] As well as elevated coal and gas costs caused by the Russia/Ukraine conflict, a '*volatile La Nina summer also played its part ... the two states* [Qld and NSW] *were also hit with lower solar production as extreme rainfall in March dampened output*'.[368]

The contents of this Chapter herald yet another public policy conflict. Not whether or not to mitigate climate change, or how best to do that, nor the dual pretexts for Wivenhoe Dam's existence. Rather the balance between ensuring housing is not constructed on lands at direct and proven risk of flooding on the one hand and the need, on the other, to maximise housing supply so as to, if not drive down prices for both renters and buyers, at least moderate them to some semblance of affordability.

The Courier-Mail editorial of 12 March 2022 analysed in Chapter 6 appeared in an edition of that newspaper which also reported

---

365 Graham Lloyd, 'Severe weather amped up climate "doomism" vote', *The Australian*, 25 May 2022, p. 13.

366 Melbourne.

367 Perry Williams, 'Price hike in power spells rise', *The Courier-Mail*, 30 April 2022, p. 60.

368 Ibid.

'[T]*enants nightmare as rents soar after floods*' due to there having been 12,800 fewer residences for rent across Queensland compared with April 2020. Hence why the asking price for a flood-free house for rent on Drynan Street in Paddington rose from $900 per week in February 2019 to $1250 by March 2022.[369] Brisbane's residential vacancy rate for the month of April 2022 was a tiny 0.7%, causing a growing number of low-income earners to face the threat of homelessness.[370] Equating to a raw number, during May, of 2403 properties in total available for residential rent, compared to 3600 houses and apartments offered to short term Air B&B style clients.[371] With a council budget sorely stretched by the post February 2022 flood repair bill, Lord Mayor Adrian Schrinner needed little incentive in his 2022/23 budget speech to levy a 50% rates premium on homes regularly listed on short-stay sites to equate their rates bills to commercial properties. Mayor Schrinner declared:

> *It is my hope, instead of paying extra, many owners of short-term rentals will now be motivated to put their homes and apartments into the longer term rental market.*[372]

Therefore to the extent that an owner of a subject property was to accept any governmental offers to buy back permanently flood affected homes, necessarily that is one less house available to buy or rent, diminishing the supply by one unit without a reciprocal replacement brought into supply. So long as there is no commensurate one unit reduction in demand, prices will inevitably rise.

The dilemma is palpable. Permanently removing 500 houses forever, as foreshadowed by the Deputy Premier of Queensland in May 2022, to add to the 112 homes in Rocklea already discontinued between 2011 and 2016, means 612 persons at least – plus their family members – added to the increasingly desperate pool of rental and

---

[369] Darren Cartwright '*Rent costs soar in the wake of record low vacancy rates and the floods making rental houses uninhabitable*' *The Courier-Mail*, 12 March 2022.

[370] Sarah Petty, 'Rental pool near empty for low-income earners', *The Australian*, 3 June 2022, p. 4.

[371] Mackenzie Scott, 'Airbnb owners slugged on rates', *The Australian*, 16 June 2022, p. 6.

[372] Ibid.

mortgage applicants in a market as competitive as Queensland. Officially the worst in Australia by year's end, with Brisbane recording the sharpest decline in affordability, and regional Queensland being declared the least affordable rental market nation wide.[373] Though with 50,000 people moving to Queensland compared to 22,000 who did so in 2019,[374] the statistic, though grim, is explicable, moving even the notoriously parochial NSW centric *Saturday Telegraph* tabloid newspaper to concede '*who wouldn't want to*' '*escape to the sunshine of Queensland ... Endless beaches, constant sunshine, what's not to like*' due to Sydney housing affordability being '*the worst it has ever been*'.[375] Basic supply and demand theory provides that decreasing the total number of private dwellings available for rent or purchase, combined with a '*wave*' of interstate migration to Queensland '*which began with Covid*',[376] will cause price rises to the extent that the consequential demand is not met elsewhere.

Speaking of the virus, a rise in Australia's national fertility rate of 1.7 babies per woman in 2021 compared to 1.59 in 2020 and 1.67 in 2019 was attributed to peak lockdowns and work from home orders.[377] Meaning that, come 2041 – 2051, yet another cohort of first home buyers and renters will need to be housed somewhere, somehow.

Especially when the lesson of 2022, compared to 2011, is that new sources of greenfield areas suitable for housing development have fallen due to the inherent riskiness of so many areas within South-East Queensland to riverine or overland flooding. A fear that did not eventuate after 2011 and if anything created too blasé an approach given, firstly, the bromide that the floods class action would 'right' the Wivenhoe 'wrong' – the wrong which the nation's highest

---

373 Samantha Healy, '"Unaffordable" the next', *The Courier-Mail*, 29 November 2022, p. 13.

374 Lydia Lynch, 'Little trickle-down from rivers of gold', *The Australian*, 2 December 2022, p. 6.

375 Editorial 'Migration Shift', *The Saturday Telegraph*, 12 November 2022, p. 76.

376 Editorial, 'Property cannot just be for the elite', *The Gold Coast Bulletin*, 6 September 2022, p. 12.

377 Melanie Burgess, Tamaryn McGregor, 'Oh baby! Town is fertile ground', *The Courier-Mail*, 14 November 2022, p. 3.

court was not prepared to entertain could have existed – and secondly a view stemming as a result of '*the 2011 event being described colloquially as a "dry" or "sunny day" flood*'.[378]

Returning to *The Weekend Australian*'s Graham Lloyd's invocation of '*recent severe flooding on the east coast* [having] *had an impact on how some people voted*'[379] in the context of the Greens' 'threepeat' of inner Brisbane House of Representatives victories, Stephen Bates, who won the seat of Brisbane from the LNP incumbent Trevor Evans, made no secret of his status as a renter himself and his desire to advocate for renters' rights in his new role:

> *Over 50 per cent of the electorate of Brisbane rents, so it's important to have that representation reflected in Parliament... me, being a renter, and my entire immediate circle of friends are renters as well*'.[380]

Throughout the entirety of Bates' being interviewed, no mention at all was made of climate change or other environmental policies for which the Greens are synonymous, and for which their leader was swift to credit the day after his party's wins in '*Greensland*':

> *Australia has granted the Greens a mandate to push the new Labor government to greater climate action and a phase-out of fossil fuels, Adam Bandt says ... the "Greensland" voting plunge is expected to add up to three seats in Brisbane to the party's lone Melbourne seat ... held by Bandt, the party leader. He said the gains were the result of a three-year strategy to target Brisbane electorates.*[381]

Rather the rental crisis, nation wide, was the sole focus of Atkin's report, hence his interview of Bates and inclusion of the Greens' $23 billion over a decade policy to create 750,000 homes for pub-

---

378 Hon Paul de Jersey, 'Brisbane City Council 2022 Flood Review', 9 May 2022, pp. 12-13.

379 Graham Lloyd, 'Severe weather amped up climate "doomism" vote', *The Australian*, 25 May 2022, p. 13.

380 Michael Atkin, ABC *7.30* telecast, 29 June 2022.

381 Mike Foley, "Greensland' gains creates mandate for end of fossil fuels: Bandt', *The Brisbane Times*, 22 May 2022.

lic and community housing and 125,000 rental properties[382]. All of those homes will have to be built somewhere though. Preferably somewhere guaranteed to be flood free and without deleterious environmental impact in the construction process. Great in theory, however presently unconstructed locations in not just South-East Queensland but Australia as a whole that meet both criteria are rare to say the least.

Super imposed over such a policy problem are planning policies and other aspects of local government decision making, such being the remit of municipal councils. Necessarily there are myriad considerations to always be considered when assessing development applications: *It's damned if you do and damned if you don't*' explained the Redland City Council Mayor when sought for comment about '*a controversial residential rezoning that could demolish vegetation and almost double the population of North Stradbroke Island*' as a result of the Queensland Government using its planning power priority over local councils to permanently allocate 249 hectares of bushland on that Island for housing development.[383] North Stradbroke Island, aka Minjeerabah, is the largest of the Moreton Bay islands, and is the second largest sand island in the world. Nor is it flood prone. So what's not to love then about the proposal? 249 hectares could house lots of people wishing to depart floodplain living once and for all. Paradoxically, given the primary content of this book, North Stradbroke's Achilles heel is fire risk; being heavily wooded it has, and always will, succumb to bushfires. As an island that is predominantly national park, trunk town water infrastructure and firefighting capacity is sorely lacking. To say nothing, speaking again of the climate change topic, of the deleterious effects that felling carbon capturing trees to create land for housing would cause. Or to koalas either, for which North Stradbroke Island is home for many. The iconic national marsupial is a threatened species. So much so that up to 140 developments that could worsen the '*koala extinction cri-*

---

382 Atkin, op. cit.

383 Hayden Johnson, 'Straddie land fire threat', *The Courier-Mail*, 30 May 2022, p. 7.

sis' are therefore required to be approved by the Federal Environment Minister Tanya Plibersek in her role as '*the ultimate decision maker on developments that affect threatened species*'.[384]

Whether or not the Queensland Government decision to have overridden the local council by approving the North Stradbroke Island proposal, despite these not-insubstantial complicating factors, was motivated by any or all of its post-February 2022 flood event announcements into alternative house building and home buyback proposals, is not known. No such link was reported. However the unavoidable truth is that all housing decisions have consequences, both in terms of approving new developments and encouraging established house owners to permanently depart floodplain locations.

What then of high-density living, such as apartment blocks? Brisbane is replete with them, though in generational terms this trend is very contemporary, Brisbane's default for centuries being suburbs of free-standing houses; 'flats', to quote the then nomenclature, being the definite exception.

That change of mindset followed urban renewal projects which commenced in the early 1990's; inner city industrial precincts like the Evans Deakin shipping yards at Kangaroo Point, the Roma Street freight railway yards and the Fortitude Valley bus depot all read the proverbial room and either ceased existence altogether or moved their operations far away from the inner city. Each establishment was replaced with luxury apartment and mixed residential and business developments: Dockside, Roma Street Parklands and Emporium respectively.

This trend then extended to the suburbs; *Tennyson Breach* chronicled exactly that, when an unlovely and long disused coal-fired power station was demolished and replaced by the Tennyson Reach apartment towers and Queensland Tennis Centre. 15 years on, the march extends even further away to suburbs as far from the central business district as Wynnum. Described, not inaccurately, as a

[384] Mike Foley, 'DA-Day: Labour faces test over koala vow', *The Sydney Morning Herald*, 31 December 2022, pp. 1 and 8.

'*sleepy*' suburb, the cat was thrown amongst the low-rise pigeons when, during May 2022, a 27-storey 275 apartment complex was submitted for development approval. 27 storeys was well in excess of the Brisbane City Council's 5-8 storey height limit for Wynnum, so it was hardly surprising that the developer claimed, somewhat provocatively, that the City Council's Wynnum neighbourhood plan was '*too old*'.[385]

South Brisbane and West End, the two suburbs immediately south of the Brisbane CBD on the other side of the Brisbane River, for generations low-rise, are now rapidly turning into clones of the Gold Coast, without the beaches, due to the vast number of apartment complexes being developed and already extant. These two suburbs are however flood-prone, having suffered during both the 2011 and 2022 events; this picture, taken from this writer's office on 25 February 2022, being day one of the three-day 'rain bomb', captures both the extremity of the rainfall that day and the massive growth in apartment construction across the Brisbane River in South Brisbane:

[385] Brayden Heselhurst, 'Way is up for Wynnum', *The Courier-Mail*, 18 May 2022, p. 20.

While this picture, in South Brisbane itself, depicts the way in which those who named this funky little alcove of wine bars and eateries 'Fish Lane' saw life unintentionally imitate art when they thereafter received patronage from actual Brisbane River fish:

Apartments may of course be capable of being constructed at levels higher than previously recorded flood levels, however neither their ground floors nor basements can be flood-proofed; basements in fact bear the brunt as gravity feeds flood waters downward, rendering impossible the ability of apartment occupiers to be immune from flooding in an ability to live sense. If electrical infrastructure necessary for apartment complexes like electrical transformers are positioned in basements rather than aloft of ground, power must be shut off for safety's sake, rendering electrically operated flood pumps inoperative. Defeating the purpose entirely. Such was the unfortunate way in which West End's Riverpoint apartment complex was planned and constructed. Adding further insult to injury, a lack of warning of imminent electrical shutdown meant its elevator could not be transported to a higher level beforehand, causing its inundation repair bill alone to reach half a million dollars.[386] A most unwelcome financial outcome for that body corporate and its ultimate funders: the lot owners (due to their body corporate insurance policy being standard, and not inclusive of flood coverage).[387]

Comparatively speaking though the residents of Riverpoint were only without power for 10 days. In South Brisbane, the 'West End Central' complex on the corner of Edmondstone and Melbourne Streets was still without mains power as long after the flood event as 26 June.[388] Jonathan Sriranganathan, Brisbane City Councillor for The Gabba ward, which includes South Brisbane and West End, visited West End Central during June 2022, and uploaded a video of his visit depicting a car destroyed by the flooding remaining in the car park, caked in baked-on river mud, numerous industrial generators enabling the commercial tenancies to trade on, yet a permanently darkened below ground car park due to electricity yet to be restored to either it or the residential component:

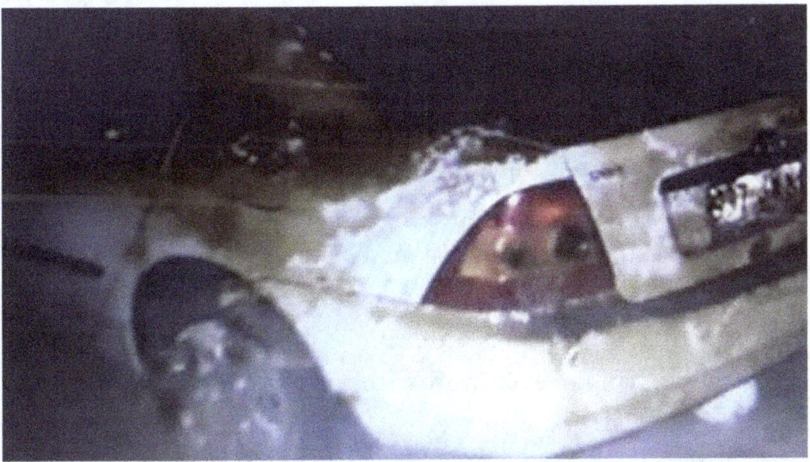

According to Cr Sriranganathan's researches, 30 of the 91 apartments were owner occupied, one half of the owners making do with

[386] Emma Pollard, *ABC News* Brisbane telecast, Sunday, 26 June 2022.

[387] Ibid.

[388] Ibid.

portable generators and walking up and down the stairs due to inoperable elevators, while the other half moved elsewhere. The remaining 60 apartments, being rentals, were all unoccupied and unoccupiable. Sriranganathan's argument, being the lack of resilience of apartment buildings to flooding even if the apartments themselves are high and dry, being made graphically apparent by his video.[389]

A similar concern was noted by Councillor Nicole Johnston whose Tennyson ward includes the perennially flood prone suburbs of Yeronga and Yeerongpilly:

> *Some parts of this City should not be built upon. Both the State and Council are allowing medium to high density building in known high risk flood areas, for example Yeronga and Yeerongpilly. The area is filled reducing flood plain storage…the basement floods, services are flooded, roads are cut … properties islanded, there is no power for a week and residents are displaced'.*[390]

The upshot being that higher density living creates higher densities of problems, and even impossibilities, due to the reciprocally higher levels of dependence apartment owners have on their common areas; losing those to flooding renders living all but untenable.

The final consideration in any chapter devoted to the property industry is the profit motive. Notwithstanding property buy backs and general exhortations not to build or buy in flood prone locations, decreases in prices occasioned by multiple flood events inevitably attract the attention of bargain hunters. '*Flooded, gutted, listed and already under offer*' is an example of a headline that conveys much more than its seven words. The article cites a three-bedroom house on a 594 square metre block in Rocklea abandoned by its owners after the February 2022 flood event; following those owners' insurers stepping into the vacuum by '*gutt*[ing] *and sanitis*[ing] *the*

[389] Jonathan Sriranganathan, Councillor for The Gabba, official Facebook page {video posted 2 June 2022}.

[390] Nicole Johnston, Submission to the de Jersey 2022 Brisbane Flood Review, 8 April 2022, p. 16.

property',[391] a Gold Coast investor paid $400,000, equating to a capital loss of $45,000 relative to its 2018 purchase price.

That fools and their money are soon parted is a truism. Also undeniably true is the rapid price growth in Brisbane's property market over the past three years. For a three bedroom house on a 594 square metre block with direct railway access to the Brisbane CBD to instead sell in 2022 for a mere $400,000 is mind boggling.

Time will only be able to tell whether that opportunistic buyer was foolish or shrewd.

Though, one month later, that buyer could be feeling rather nervous; by the end of June 2022 Rocklea residents classed their streets as '*ghost streets*' due to dozens of homes having been left to rot, doors barred up or taped off and seemingly abandoned by their flooded owners: '*It's been silent. Eerily quiet*' observed a resident of Corella Street Rocklea, referring to five properties in that street alone having been left abandoned.[392] To the extent that such a depressing trend continues, compounded by the Reserve Bank of Australia's multiple interest rate rises during 2022, any capital gain on the Gold Coast investor's $400,000 acquisition price will be years, not months, away.

Suncorp CEO Johnson's concern not to build so as to put '*people in harm's way*' is of course an entirely rational one, especially in light of Morgan Stanley analysis that his company would need to lift its natural catastrophe budgets '*substantially*' after having come in under budget twice in the past 14 years.[393] That having been said, not building so as to put '*people in harm's way*' will of itself have many law of unintended policy consequences for not just governments to grapple with but property developers also, who, incidentally, seek from financier entities such as Johnson's loan funds to finance property developments, and in turn pay those very lenders interest on

---

391 Samantha Healy, 'Flooded, gutted, listed and already under offer', *The Courier-Mail*, 27 May 2022, p. 11.

392 Matty Holdsworth, Natasha Innes, 'Flood suburbs our new ghost town', *The Courier-Mail*, 23 June 2022, p. 14.

393 Chloe Whelan, 'Suncorp expects claims surge after floods', *The Australian*, 5 July 2022, p. 20.

such lendings. Culminating in the bank Johnson helms winning a 'business bank of the year' award four years running.[394]

Hence the reference in the title to this chapter to circularity; in a rapidly population growing city on a floodplain like Brisbane, planning debates and proposals will carry so many unintended consequences that a circle rather than a continuum would be the only accurate way to depict them all.

[394] 'Business Bank of the Year Suncorp', *Money Magazine*, July 2021, p. 43.

9

May Mayhem

As stated in Chapter 6 the first week of March 2022 was, fortuitously, quite sunny, which meant that Seqwater releases from Wivenhoe Dam were able to be made so as to have fully emptied the flood storage compartment under the Manual's Drain Down Strategy.

The rest of March and all of April were unexceptional rainfall wise.

However the change of month to May 2022 introduced what became a uniform heavy rainfall forecast as a result of the formation of a '*coastal trough and inland trough*' interacting with cold air moving in a westerly direction merging with '*humid air being drawn in from the Coral Sea*'.[395]

The '*humid air*' aspect of that report was definitely accurate; the humidity Brisbane suffered through was oppresive. By way of example was the night of Sunday/morning of Monday 15/16 May: Brisbane's lowest minimum temperature was 21.8 centigrade, a full 8.8 degrees greater than the long term May minimum average of 13.0 degrees.[396] This was due to the enervating effect of the vast amount of water vapour in a saturated atmosphere, with '*almost 100 per cent humidity, higher than Cairns' current humidity levels of 60 per cent. "This humidity outbreak is like ... January February...It's ideal for mould*"' explained a Bureau of Meteorology employee.[397]

Ideal for mould and ideal for mould-removal contractors too: one in five Queensland homes were invaded by mould at an average clean-up bill per household of $500.[398]

---

[395] Felicity Ripper, 'Threat of 500mm deluge', *The Courier-Mail*, 10 May 2022, p. 5.

[396] Bureau of Meteorology Brisbane, May 2022, Daily Weather Observations.

[397] Matty Holdsworth, 'Humidity swamps Brisbane', *The Courier-Mail*, 17 May 2022, p. 16.

[398] Matty Holdsworth, 'One in five fight mould', *The Courier-Mail*, 26 May 2022, p. 19.

A terrible outcome for renters though who, despite having lawful grounds to summarily terminate their leases of mould-afflicted properties, simply could not do so due to the paucity of affordable alternatives. A Northgate renter who had made almost 30 complaints to his letting agency about a mould outbreak assessed by an expert as rendering his rental '*unliveable*', with consequential suffering of breathing issues adding insult to injury, was nonetheless unable to up and leave such an untenable residence due to a rental vacancy rate of less than 1%.[399]

Even cockroaches came out to celebrate; a '*dramatic increase*' in first-stage cockroaches emerging from eggs early instead of waiting out the winter months was recorded by the Queensland Museum's curator of entomology.[400]

The humidity was not much fun either for those queuing up to pre-poll for the 2022 Federal Election, with Australian Electoral Commission staff having to act to enable vulnerable voters to jump queues that had been so long as to have caused, at their worst, 90 minute wait times.[401]

Homeowners similarly suffered in unforeseen ways; the high moisture content of the air during May caused wooden doors and windows to expand, making them difficult, if not impossible, to open and close. Especially a negative for those heeding advice to open doors and windows to allow more natural airflows in a bid to defeat the mould outbreaks. Brisbane radio personality, builder and former Wallaby fullback, Greg Martin, explained the way in which wooden floorboards, being a staple choice for thousands of Queensland homeowners, warped as a result of the high moisture contents, causing resulting difficulties to those unfortunate enough to have had their floors installed by contractors who used bullet headed/straight shafted nails on installation rather than screwshank

---

399 Matty Holdsworth, 'Humidity swamps Brisbane', *The Courier-Mail*, 17 May 2022, p. 16.
400 'Roach plague hits city', *The Courier-Mail*, 20 May 2022, p. 5.
401 'Long Waits, Big Lines', *The Courier-Mail*, 17 May 2022, p. 9.

nails better able to restrain the warping effect.[402] An expert flooring contractor who phoned into Martin's station cited the causes this way:

> *We've had such a dry period … 2017, 2018, 2019 … people were laying lots and lots of timber floors, then we get all the rain … all those floors grow and … skirting boards are only 11mms thick these days in general and you haven't got a lot of room for expansion and you've got to put in expansion and there's been a few cowboys out there who have been relying on the dry weather to just slap floors in and now suddenly they are all pressed up against the walls and jumping off the floor.*'[403]

Then there were the many householders – owners and renters alike – who, like myself were responsible for washing kids' school uniforms, duly challenged and plaintively looking to every possible internal location to fashion a form of internal clothesline option, then wait literally days and days to do what 10 minutes in the summer sun on the outside line can and always did. '*Even when it's not raining, we are struggling to keep the dampness out of the house*' was the lament of one West End renter forced to dispose of thousands of dollars worth of ruined clothes and furniture.[404]

May 2022 was as frustrating a month as a strange one, wholly unlike any previous others. South-East Queensland averaged 3.1 hours of sunshine per day during May 2022, on a par with Torshavn in the far from tropical Baltic nation of Denmark, hence the clever headline '*South-East's rain as sun king is over*'.[405]

As if renting accommodation and washing clothes was not difficult enough, another negative outcome in the month of May was a consequential rise in prices of essential fruits and vegetables due to flooding in prime farming lands. Salad green crops ripe for picking before May 2022 were waterlogged, causing cashflow problems and

---

402 *Triple M* Brisbane, Big Breakfast broadcast, 1 June 2022.

403 *Triple M* Brisbane, Big Breakfast broadcast, 1 June 2022.

404 Matty Holdsworth, 'One in five fight mould', *The Courier-Mail*, 26 May 2022, p. 19.

405 Tricia Rivera, 'South-East's rain as sun king is over', *The Courier-Mail*, 24 May 2022, p. 3.

very large losses in the millions of dollars for farms in the Lockyer Valley west of Brisbane, the Lockyer Valley being an iconic 'food bowl' for its cultivation of various grains, vegetables and breeds of livestock.[406] So much so that the Lockyer Valley supplies up to 70% of winter vegetables supplied to Australia's eastern seaboard.[407] Notably the Lockyer Valley is transited by the Lockyer Creek mentioned earlier in this work as a contributing sub-catchment to the overall Brisbane River system, extensively flooded in January 2011 and February 2022, and during May 2022 as well. Heavy rain and low sunlight '*in Queensland*' being expressly cited by *The Daily Telegraph* in an article advising its Sydney based readers of supply and quality reductions of zucchini, beans, gourmet and cherry tomatoes and broccolini.[408] Inflation, reinstated to public consciousness as people previously locked down by Corona Virus began spending again, to say nothing of the oil price shock caused by the Russia/Ukraine conflict, was also forecast to grow further due to the fresh food shortages inevitably causing the fresh food inflation rate to outpace that of dried grocery goods, and, to compound the miseries, increasing pressure on the Reserve Bank to enter into a prolonged period of interest rate rises.[409]

Lettuces, normally an unremarkable staple food thrown into refrigerator crispers without a second thought, suddenly roared into public consciousness due to lettuce crops across the Lockyer Valley being ruined, constraining supplies and '*drastically increas*[ing]'[410] purchase costs at consumer levels.

---

[406] Deloitte Access Economics & Queensland Reconstruction Authority, 'The social, financial and economic costs of the 2022 South-East Queensland Rainfall and Flooding Event', June 2022, p. 24.

[407] Georgie Hewson, *ABC News* Queensland telecast, 15 May 2022.

[408] Angie Raphael, 'Brace for more bill shock: Coles CEO', *The Daily Telegraph*, 3 June 2002, p. 21.

[409] Eli Greenblat, 'Shoppers warm to frozen food in vegie drought', *The Weekend Australian*, 28-29 May 2022, p. 19.

[410] Deloitte Access Economics & Queensland Reconstruction Authority, 'The social, financial and economic costs of the 2022 South-East Queensland Rainfall and Flooding Event', June 2022, p. 24.

The May minor flood event caused Brisbane's May 2022 rainfall to exceed 110mm by the afternoon of Thursday 12 May 2022; not even 13 days into the month of May Brisbane exceeded the May 2015 total, 2015 being the last year in which the month of May attained that figure. Even then May 2015 required an entire month to reach that.[411] Brisbane's annual average rainfall total of 1011.5mm was caught and well and truly surpassed by 18 May 2022 when a total of 1486mm was reached.[412] With catchments sodden due to 1.3 metres of rain in total by not even the 6th month of the year 2022,[413] it was not particularly surprising that the Brisbane River, at 22:49 on 17 May 2022, reached 1.81m in height at the City Gauge, 11 centimetres over the official minor flood level of 1.70m, due to rainfalls which would otherwise have been absorbed running straight off into the River.

Thankfully without the loss and damage of 2½ months earlier, which at 3.85m exceeded major flood level, although the loss of two lives nonetheless proves that there's nothing 'minor' about any weather event which claims a life.

Another concomitant effect of prolonged rains and saturated soils was a spate of tree fellings. On the last day of the month, 31 May, the weather reversed completely; an icy polar blast that caused southern Australia to shiver produced low humidity and gusty winds to replace, for the final day of the month anyway, the dominant paradigm. Trees formerly anchored in hard dry soils found themselves unable to withstand the high winds as their much flimsier and muddier foundations buckled, causing multiple tree toppling overs.

Finally, when it seems no more mayhem could possibly be occasioned in a single month, came one last twist. Also on 31 May, an earthquake shook the ground ever so slightly and shortly at 1.25pm,

---

411 Felicity Ripper, Matthew Johnston, 'The snarling floods of May', *The Courier-Mail*, 13 May 2022, p. 10.

412 Felicity Ripper, 'One year's worth of rain in five months', *The Courier-Mail*, 18 May 2022, p. 7.

413 Felicity Ripper, Matthew Johnston, 'The snarling floods of May', *The Courier-Mail*, 13 May 2022, p. 10.

when a five second long magnitude 2.9 earthquake centred on Brendale, in Brisbane's outer north-west, saw the month out with a shake.[414] Thankfully damage free, though as a metaphor, a singularly apt one.

Wivenhoe Dam management began with a 10 May 2022 forecast from Seqwater that '[M]*inor gated releases from Somerset, Wivenhoe and North Pine Dams are possible in the next 12 to 48 hours due to BOM* [Bureau of Meteorology] *rain forecast*'.[415]

Thankfully the forecasts proved correct and the waters released were not forfeited.

Akin to February, the May 2022 event also caused large amounts to have fallen in the Lockyer Valley and the Bremer River catchment areas, causing Seqwater to have announced on 13 May 2022:

> *Releases from Wivenhoe Dam have temporarily stopped, as inflows to the Brisbane River from Lockyer Creek and the Bremer River increase due to ongoing rainfall. The Lockyer Creek and Bremer River flow into the Brisbane River downstream of Wivenhoe Dam. The benefit of Wivenhoe is being able to store water in the dam whilst the Bremer and Lockyer flows peak, ultimately reducing the height of the Brisbane River. At the moment, we are utilising a relatively small amount of the flood storage compartment and have approximately 90% available to respond.*[416]

The strategy worked; releases from Wivenhoe Dam restarted at 3:30am on 14 May 2022 after inflows to the Brisbane River from the Lockyer Creek and Bremer River generally peaked,[417] continuing to Sunday 22 May 2022 to empty the flood storage compartment[418]. The highest the Brisbane River reached at the City Gauge was 1.81m, surpassing the minor flood level by 11 centimetres, proving,

---

414 Nathan Greaves, 'Afternoon earthquake rocks region', *The Somerset*, 8 June 2022, p. 7.

415 Twitter @seqwater 10 May 2022.

416 Twitter @seqwater 13 May 2022.

417 Twitter @seqwater 14 May 2022.

418 Seqwater Facebook page, 20 May 2022.

yet again, the unavoidable impact on the city of the 48.416% of its catchment areas downstream of and independent of Somerset and Wivenhoe Dams.

May mayhem also caused upheaval in at least one corporate boardroom: that of Sydney based ASX listed property giant: Mirvac. Life in a floodplain city has not been kind to Mirvac; *Tennyson Breach* listed out the ways in which a wildly financially successful 2007 launch by that company of off-the-plan contracts for a complex of six luxury apartment buildings on the banks of the Brisbane River at Tennyson turned very sour very quickly as a result, firstly, of the Global Financial Crisis of 2008/2009 and then the January 2011 flooding which so devastated the then complex of three (the future planned three apartment blocks not going ahead), culminating in Mirvac's 2011 announcement to the Australian Stock Exchange of writing off the entire value of all of its unsold apartment stock at Tennyson[419].

Eleven years later Mirvac was battered yet again by a Brisbane flood. The differences being that this time rather than a bespoke apartment complex constructed by it, the asset Mirvac lost was a shopping centre purchased by the company after being constructed as long ago as 1967: the Toombul Shopping Centre (following page). The second difference was that the flooding was not of the Brisbane River. Rather the wholly separate Kedron Brook, the forces of which were analysed in Chapter 2 and analogous, on 27 February 2022, to a category 5 cyclone. On that day Kedron Brook's comparatively low lying northern bank at Toombul, conveying the vast amounts of floodwaters as depicted in that chapter's photographs, caused those waters to divert leftwards directly, firstly to devastate the car park and then inside the shopping centre itself:[420]

Though Mirvac's initial position, post-February 2022, was to spend six post-flood months rehabilitating the centre, during May

---

419 *Tennyson Breach*, pp. 106 & 107.

420 *Your Neighbourhood Magazine*, 'Mirvac & Council Discuss Future – Toombul Shopping Centre', 7 June 2022.

2022's encore rain event {albeit one in which Kedron Brook kept within its bounds}, Mirvac chose to throw the towel in completely, announcing on 18 May 2022 that the centre was too severely damaged to repair and reinstate to its pre February 2022 position.[421] 55 years of shopping history was summarily terminated. All leases cancelled. A very grave corporate decision for Mirvac to have made. Up to 130 businesses, ranging from national to small operators, were advised of the permanent closure, in the style of the Corona Virus era, firstly by email and secondly by Zoom, with Mirvac reported to have offered tenants '*thanks*',[422] and directions to government grants and mental health support,[423] but, as of the announcement, no compensation.

That position had thankfully changed by the next month: '*Mirvac has announced a multimillion-dollar support package* [which] … *would include a cash payout equivalent to three months rent… on top of previously announced support measures including waiving all debts*

---

[421] Glen Norris and Chris Herde, 'End of era as floods force centre to close', *The Courier-Mail*, 19 May 2022, p. 12.

[422] Glen Norris, Chris Herde, 'Down the drain Toombul firms reel after shock closure', *The Courier-Mail*, 20 May 2022, p. 24.

[423] Glen Norris, Chris Herde, Samantha Scott and Brendan O'Malley, 'Tenants lose it all as Toombul sinks', *The Courier-Mail*, 26 May 2022, p. 12.

incurred during Covid-19 ... providing free storage to store stock and assisting with new fitouts at new locations'.[424]

Shoppers too noticed the consequential inconvenience; 1700 car parking spaces for shoppers had been removed in the meantime as a result of Toombul's intended interim closure, causing a '*parking nightmare*' at the closest alternative shopping centres – Chermside, Stafford and Lutwyche: '*... like Christmas rush every weekend there* [Chermside] *now*' theorised QUT retail specialist Gary Mortimer.[425] Christmas rush definitely came early to Chermside, and also outer Brisbane's North Lakes shopping centre, during the fathers' day weekend of early September. Drivers seeking to exit Chermside's car park faced delays of up to 3 hours and 75 minutes at North Lakes, with Toombul's closure and the consequential need for 1700 extra cars to find alternative shopping centres being directly cited as the cause.[426]

Incidentally the Stafford shopping centre Mortimer spoke of is also situated alongside the same Kedron Brook which devastated Toombul out of existence.

Due however to a fortuitous topographical and watercourse directional difference, Stafford Shopping Centre was entirely untouched during February 2022. This self-taken picture is from the Stafford Centre's south-western corner carpark looking west into Alderley. The image depicts the Brisbane City Council's undertaking of urgent restoration works to the bikeway spur off the Kedron Brook bikeway directly into the Stafford Centre's car park, including replacement of yet another pedestrian over-bridge which was severed by the torrent on 27 February 2022:

424 Glen Norris, Chris Herde, Madura McCormack, 'Toombul rescue package', *The Courier-Mail*, 10 June 2022, p. 24.

425 Glen Norris, Chris Herde and Brendan O'Malley, 'Shopping centre car park "like a crab pot"', *The Courier-Mail*, 24 May 2022, p. 2.

426 Jordan Bissell, *7 News* Brisbane telecast, 5 September 2022.

One savvy Toombul retailer prudently took out a flood insurance policy, however 95% of Toombul's tenants did not. In a case of tragic timing, Toombul's gymnasium business had spent more than $3 million in fit out costs during 2021[427] which, as a result of the closure announcement, was $3 million thrown away, save only from what could be clawed back from the compensation package.

Of course one can not judge those businesses harshly; how many of those retailers would have expected that 1007mm of rain would have fallen in Alderley,[428] a suburb not particularly close to Toombul, and with no other connection save only for Kedron Brook,

[427] Glen Norris, Chris Herde, Samantha Scott and Brendan O'Malley, 'Tenants lose it all as Toombul sinks', *The Courier-Mail*, 26 May 2022, p. 12.

[428] Supra, fn 13.

which, on 27 February 2022, was so catastrophically overloaded with floodwaters that its comparatively low lying bank at Toombul caused those waters to divert leftwards directly, and fatefully, inside the Toombul centre.

Especially when, in an otherwise broadly flood ravaged and damaged city, the majority of the major shopping centres were unaffected.

Build an asset in Queensland or purchase one; either way the Sydney based Mirvac could be readily excused for wanting to withdraw from this state altogether.

10

Conclusions

Chapter 1 began with my grandmother's 1910 recitation of '*Rain rain go away*', and Chapter 10 begins with the exact same song being sung in 2022:

Some things never change.

Instagram is a device which those alive in 1910 were unable to enjoy, however the poster's frustrations were just as evident and identical in 2022 as 1910.

What I recalled of my grandmother's cautionary tale to be careful what one wishes for came rushing starkly back in February 2022 upon reading dbayliving's post. Casting another look back at Chapter 1's 1900-2010 rainfall graph vividly depicts the way in which, in the year after my grandmother and her friends succeeded in persuading rainfalls to leave Plainland, 1911 duly delivered; it was a year notably bereft of rain.

Who is to say 2023 won't be the same in and around South-East Queensland?

2023 incidentally marks the bicentenary of John Oxley's reconnaissance of the then un-named Brisbane River. John Oxley was instructed by the Governor of the colony of New South Wales Sir Thomas Brisbane to travel north to assess new sites for penal settlements. Oxley's legacy lives on by having the Brisbane River's Oxley Creek sub-catchment, and cognate suburb of Oxley, both named after him. Between 1 and 5 December 1823, Oxley and his assistants duly traversed both the modern day Brisbane and North Pine Rivers, culminating in a 6 December 1823 return to Sydney with associated recommendation to establish a Moreton Bay settlement.[429] Matters proceeded apace; Oxley made repeat voyages, Governor Brisbane made his first trip north from Sydney in November 1824, and the first official exports from the then nascent '*Brisbane Town*' – being pine logs – began in 1825.[430] By the time the 100th year after Oxley first encountered the Brisbane River arrived, '*general and individual wellbeing among all classes mark*[ed] ... *the widespread prosperity of Queensland at the close of 1922*', so editorialised the then Brisbane *Telegraph*, along with a prediction of a '*splendid season that now awaits us after the menace of general disaster from drought a few weeks ago*.'[431]

The reference to '*the menace of ... drought*' proving consistent, yet again, with Chapter 1's thesis that drought and Brisbane go hand in hand, even though the common conception is that the polar opposite of drought truly defines Brisbane.

Even more noteworthy was that the epithet '*menace*' was used to describe a drought only 29 years after the '*Brisbane River turn*[ed]

[429] J. G. Steele, *The Explorers of the Moreton Bay District*, University of Qld Press, 1972, p. 87.

[430] Ibid, p. 91.

[431] Dot Whittington, 'Optimism reigned in the wake of war', *The Sunday Mail*, 1 January 2023, p. 53.

from a harmless servant into a devouring monster' creating a '*new inland sea ... swallowing buildings, schools and bridges*'.[432] Clearly the 1893 flood event, being the second highest in Brisbane's eye-witnessed history, was all of menace and more, yet by definition floods depart almost as quickly as they arrive. Whereas the latency, sometimes decades long, of droughts, means that the despair remains ongoing, unable to be hosed away like mud off walls after flood peaks subside.

While on the topic of forgetting – perhaps even wilful forgetting – Dr Cook's researches in River note that the 8.43 metre Brisbane River flood of 1841 occurred '*before Brisbane's first land sales had been held*'.[433] '*Moreton Bay opened for free settlement in 1842*',[434] however the 1841 flood '*went largely unreported in both the Sydney and Moreton Bay press. Rivers offered the promised land of untapped economic potential; accounts of floods did not sit comfortably with the prevailing narrative of prosperity and progress*'.[435]

The result being that, by 1893, more than half a century later, the epic flood of that year so shocked and surprised the then population of Brisbane. A population that had been warned of the impending doom – '*Henry Plantagenet Somerset ... had desperately, but unsuccessfully, attempted to warn Brisbane of the approach of the 1893 flood*'[436] – yet chose not to take heed.

The economic potential rivers offered during the 1800's cited by Dr Cook was multiplied in the first half of the 20th century as advances in engineering enabled dam projects to commence; '*hailed as providing jobs, water security, opening up additional arable land or for other industrial uses ...* [dams] *were commonly seen as substantially improving the public good*'.[437]

---

[432] Extracted from the 'Australia 1893' page of *Australia Through Time*, 2002 Edition, Random House Australia Pty Ltd, p. 114.

[433] *River*, p. 10.

[434] Ibid.

[435] *River*, p. 12.

[436] Supra, fn 92.

[437] Bruce Kingston, 'The sad history of water politics in SE Qld and the bad politics that led to the 'solution' of desalination', Case Study 1, *White Elephant Stampede* [Eds: David Gration, Bruce Kingston and Scott Prasser], Connor Court Publishing, 2022, p. 31.

Consistently with the '*public good*' wisdom of rivers enabling riverside settlements, John Oxley's contributions of 1823 and 1824 were memorialised 100 years later. The resulting structure still exists today, though even long-term Brisbane residents could be excused for being completely unaware of its existence in its extremely blink and you will miss it location on Brisbane's North Quay, across the road from its intersection with Makerston Street:

Blink and you will miss it due to a combination of the overgrown nature of the vines around the memorial and its unfortunate location literally in the middle of what is effectively the commencement – heading south – and end – heading north – of the Riverside Expressway which ultimately forms the M1 linking Brisbane to the Gold Coast. The roar of multiple vehicles driving by on both sides and the lack of safety for persons wishing to attend likely explains why this monument is barely known amongst the population. Also the plaque's reference to Oxley's 'discovery' inappropriately ignores the many thousands of years of stewardship of the Brisbane River by the Turrbal people, who are the traditional owners and custodians of what is now Brisbane, stretching from Elimbah in the north, the Logan River in the south and Moggill in the west,[438] and the Jagera people westwards of Moggill. Such ignorance being nonetheless par for the course during September 1924, when the stone was first struck so as to signify the centenary of establishment of a formalised township on what is now Brisbane's Central Business District.[439]

The '*dbay*' of dbayliving's Instagram post which introduced this chapter is shorthand for Deception Bay, a coastal community midway between Brisbane and the Sunshine Coast on the banks of Deception Bay, the northern-most of Moreton Bay's various sub-bays. So named, by John Oxley actually, as a result of Oxley being deceived by Deception Bay's shallow nature, thinking it instead to have been a river.[440] This picture, originally intended to depict the very non-hereditary nimbleness of your present writer's son, was taken from the suburb of Deception Bay looking eastward. Demonstrating both Deception Bay's shallowness at low tide, and also it's bay within a bay nature: Scarborough, being the Redcliffe peninsula's northern most suburb, is visible in the background, with the broader Moreton Bay being to Scarborough's north and east:

---

[438] https://www.turrbal.com.au/

[439] 'John Oxley Memorial Stone, 'Unknown Pioneers Monument & Riverside Expressway Opening Plaque', Brisbane City Council Heritage Information page, May 2011.

[440] 'Deception Bay – bay (entry 9563)', *Queensland Place Names*, Queensland Government, 31 October 2022.

As such Deception Bay's direct proximity to coastal waters means, as a locality, that its prospects of regular rainfalls are reasonably high. Whereas my grandmother's childhood location of Plainland situated roughly halfway between Brisbane and Toowoomba and broadly co-extensively, in a longitudinal sense to Wivenhoe Dam, suffers from the empirically proven inability of rainfalls to penetrate inland nearly as regularly as they fall on coastal locations.[441]

An apt illustrator was the way in which Gatton, a town a mere 10 minute or so drive west of Plainland, and the effective commercial centre for the Lockyer Valley, recorded its highest ever daily rainfall total of 302mm on February 26, with its total between 25 and 28 February 2022 amounting to 531mm[442]. A genuine record total to be sure, and fully consistent with the extremely rapid rise in the levels of Wivenhoe Dam – to Gatton's north-east – during the February 2022 flood event, yet still 145.8mm lower than the three-day total of 676.8mm which Brisbane received. Therefore while those living coastally might be excused for wishing rains away, those living inland would be aware that both the frequency and regularity

441 Peter Davie, Errol Stock and Darryl Low Choy (eds), *The Brisbane River: A sourcebook for the future*, The Australian Littoral Society, 1990, p. 3.

442 Deloitte Access Economics & Queensland Reconstruction Authority, 'The social, financial and economic costs of the 2022 South-East Queensland Rainfall and Flooding Event', June 2022, p. 24.

of rainfall totals commensurately and not insubstantially decreases the further inland from Moreton Bay one ventures.

Governments are just as equally aware – or should be – as they bear the budgetary brunt of policies and procedures designed to counteract diminished water supplies as a result of a thirsty greater Brisbane voraciously consuming Somerset and Wivenhoe Dam waters whilst inflows dry up due to their catchments' westerly non-coastal locations:

> *... while lowering the full supply volume to supplement the flood mitigation volume did indeed result in an overall lowering of flood damage costs, this was counterbalanced by an increase in the costs associated with water supply, including the increased frequency of implementation of water restrictions, operation of desalination and recycling plants, and bring-forward of water supply augmentation plans.*[443]

It is for these reasons that your present writer maintains his Chapter 6 contention against pre-emptive '*massive*' releases of discharges out of Wivenhoe Dam.

Or even managed 90.5 to 80 percent releases, as occurred on the weekend of 15-16 October 2022.

Moreover, this contention remains pressed notwithstanding the 16 August 2022 double-negative La Nina and Indian Ocean Dipole forecast and eerie coincidences with 1974, 2010 and 2021. Firstly because, some two months after the pre-emptive 90.5 to 80 percent release from Wivenhoe Dam, the reporting of August 2022 was suddenly and rapidly walked back:

> *After three successive La Niñas, there are strong signs that the latest La Niña is quickly waning and will release its grip and move towards neutral in its effect on Australia in the second half of summer into autumn ... This means we can expect an end to the relentlessly wet weather which has lashed large parts*

[443] Dr Luke Toombes and Rob Ayre, *Resetting Expectations of Brisbane's Flood Protection*, p. 10.

of Australia in recent summers, and for much of 2022 – with a return to drier, potentially warmer conditions as we move through summer.'[444]

Weatherzone went on to explain and illustrate how an ocean pattern named the Kelvin wave, which moves across the Pacific underneath the ocean surface, had commenced transporting warmer-than-usual Australian waters back into the central Pacific.

'*Because warm-than-usual* [sic] *water near Australia is La Niña's fuel*',[445] to the extent of such transportation away the balance will start to tilt back towards a hotter and drier overall weather paradigm.

So much so that by year's end a declaration was made by a *Courier-Mail* sub-editor '*It's adios La Nina*' to headline a report that '[I]*nitial predictions indicated Queensland would experience higher than average rainfall and possible flooding during December, however most areas received less than average monthly totals ...*', and further going so far as to cite a possibility of a complete u-turn to an El Nino phase by May 2023.[446] El Nino being the name given to La Nina's opposite cycle, meaning that instead of warmer Pacific waters, colder waters are diverted to Australia's east coast which in turn lead to less evaporation, lower cloud cover and drier conditions overall contributing to increased risks of droughts, heatwaves and bushfires.[447]

Quite a radical divergence of messaging during one single 4½ month period – from the Premier's foreboding, '*I don't want Queenslanders to be alarmed, but ...*', wording of 16 August, to 31 December 2022's potential for the reinstatement of dry and drought like conditions.

---

444 Anthony Sharwood, 'Strong signs that La Nina is breaking down', Weatherzone.com.au, 20 December 2022 12:33 PM AEST.

445 Ibid.

446 Taylah Fellows, 'It's adios La Nina', *The Courier-Mail*, 31 December 2022, p. 13.

447 Mike Foley, 'End to big wet on the horizon', *The Sydney Morning Herald*, 5 January 2023, p. 5.

Secondly even if heavy rains are repeated during the summer of 2022-23, notwithstanding the alteration in the Kelvin wave, the combined land mass of Queensland and New South Wales is very large; there is no guarantee that the areas of both states so brutally devasted in February and March 2022 will be hit twice. '[T]*orrential rain … smash*[ed] *the Whitsunday region … at Airlie Beach, Proserpine and Bowen*' between 14 and 17 January 2023, with nearby Hamilton Island recording more than twice its entire month of January average rainfall by the midpoint of that month,[448] yet it was sunny skies which prevailed day after day over the Somerset and Wivenhoe catchments as the new year emerged.

Thirdly and finally, pre-emptive Wivenhoe Dam releases could in fact worsen, rather than ameliorate, flooding in Brisbane; Toombes and Ayre describe as a '*significant limitation of current rainfall forecast*[ing] … *the coarseness of the spatial definition and the unquantified uncertainty that may be associated with it. If a release is made in response to forecast rainfall that subsequently falls in the catchments downstream of the dam rather than upstream … then the release may exacerbate flooding …*'[449]

As 2022 in fact proved: loss and damage to numerous suburbs in Brisbane occurred despite a paucity of actual Wivenhoe Dam intentional outflows during the month of February.

To say nothing also of the other vast risk of pre-emptive releases: the basal, albeit inconvenient, truth that waters released from Brisbane's primary water storage reservoirs Somerset and Wivenhoe in pursuance of such a pretext will definitely not somehow be restored if the forecasts are wrong and replacement rains fail to fall over them and their upstream catchments.

A similar theme of extreme rain being replaced by just as extreme drought was voiced by Brisbane Lord Mayor Adrian Schrinner dur-

---

448 Fia Walsh, Duncan Evans, Chloe Grimshaw and Janessa Ekert, 'Water Torture Will Get Worse', *The Courier-Mail*, 17 January 2023, pp. 8 & 9.

449 Dr Luke Toombes and Rob Ayre, *Resetting Expectations of Brisbane's Flood Protection*, p. 11.

ing May 2022 in an opinion piece penned for the *Courier-Mail* under the sharply worded headline '*Drought could kill the Games if we don't act*'.

As a starting proposition, the predominantly outdoor nature of summer Olympic events would be more conducive to high-quality performances in sunnier rather than rainy weather; if ever two weeks of fine weather could somehow be locked in, the two Olympic weeks during August 2032 would be perfect.

However Schrinner's point was not that; rather it's his '*nightmare scenario*' of 2032 arriving whilst Brisbane is in the '*grip of a really bad dry spell*'.[450] The theme of Chapter 1, as proven by the graph of 1900-2010 rainfalls, is that South-East Queensland is inherently drought prone with a history of long and depressing dry spells. Yes the flood events, when they occur, are spectacular in effect, and, to be sure, the underlying proposition to Chapter 8 is that they are increasing in extent and frequency. However it remains submitted that the long term default of South-East Queensland being drier rather than wetter will continue; 2022 being an upward blip on a graph just as 2010-2011 was.

Bringing back into focus the dual purpose of Wivenhoe Dam, as made clear during its gestational period when river flooding had long faded from public and bureaucratic consciousness: reliable drinking water supply for a city that, by 2032, could well be home to one million more people than in 2022 as a result of '*our enviable weather, relaxed lifestyle and excellent job opportunities, thanks to a city booming off the back of the Olympics*'.[451]

Schrinner's population forecast has verifiable backing; '*Queensland's population is growing by one person every five minutes and 37 seconds*'.[452] Consistently, 426,000 new dwellings were con-

---

[450] Adrian Schrinner, 'Drought could kill the Games if we don't act', *The Courier-Mail*, 10 May 2022, p. 33.

[451] Ibid.

[452] Gert-Jan de Graaff, Brisbane Airport Corporation CEO, 'Unity is key to success', *The Courier-Mail*, 13 November 2022, p. 13.

structed throughout Queensland between 2012 and 2022, yet investment in major water storage and other infrastructure has stagnated.[453]

Observing correctly that for each of the past three summers '*storage levels across our dams dropped below 60 per cent*',[454] Mayor Schrinner argued:

> [W]*hether it's a dam or desalination or … recycled water, this is a conversation that South-East Queensland has to have*' because '[S]*ure, it might be raining today. It may not be in 2032.*[455]

These such '*conversations*' will need to begin, difficult as they will be. Difficulties which will only be heightened by the concomitant public policy price of buying back houses in flood-prone areas: one less house means one less unit of supply in a context of already unaffordable housing prices, which effect will only compound as the '*one million more*' phenomenon Schrinner and de Graaff separately identified inevitably emerges in the lead up to the 2032 Olympic games.

Queensland's '… *stellar population growth* … [is] *driven* … *by interstate migration and further pushed along* … *by reopening international borders*' because '*Brisbane remains affordable. This further exacerbates population movement given affordability challenges elsewhere but is also encouraging a high level of interstate investor activity. For $1 million, you can buy in Sydney's highest crime suburb of Blacktown. For the same price, you can buy in Brisbane's Kelvin Grove*',[456] Kelvin Grove being literally in walking distance to the Brisbane CBD. Doing so from Blacktown to Sydney's CBD being impossible in anything less than eight hours, so Ray White's juxtaposition of what identical values can buy was done quite vividly.

Or, alternatively, one could construe those figures in a June 2022

---

[453] Hayden Johnson, 'Premier, it's time for an informed discussion', *The Sunday Mail*, 14 August 2022, p. 9.

[454] Adrian Schrinner, '*Drought could kill the Games if we don't act*', *The Courier-Mail*, 10 May 2022, p. 33.

[455] Ibid.

[456] Ray White, *Now*, February 2022, p. 15.

context of a single house in Blacktown being worth 2.5 houses in Rocklea, despite the former suburb being 22 railway stations away from the Sydney CBD, with commensurate longer total travel time, compared to Rocklea being only 10 stations away from Brisbane's Central station.

The 2032 Olympics could however, in an indirect way, assist in a planning sense given the proposal to resume obsolete industrial land along the northern Brisbane River bank at Hamilton Reach in Brisbane's north-east to create the Brisbane 2032 Olympic and Paralympic Games Athletes Village.[457] Hamilton Reach is only kilometres away from the Brisbane River mouth, and, as a result of the River being so much wider downstream than it is in and around central Brisbane, the River there possesses sufficient capacity to convey rather than spread floodwaters. Tellingly no adverse affection was caused to Hamilton Reach's river banks during both of 2011 and 2022.[458]

This being the case, repurposing those banks away from industry to high density residential living is not only logical, it should be prioritised by all Brisbane City Councils including and subsequent to that of Schrinner's given the genuine paucity of truly flood-free land in and around Brisbane capable of being repurposed for residential living.

Not only that, there is unique scope for the Federal Government, not a tier of government known for relevance in the planning space, to step in; the consistent narrative of the newly elected Albanese government was the way in which it had 'inherited' nine years of alleged shortcomings from the three previous iterations of the Coalition government between 2013 and 2022. One of those 'inheritances', and a tangible one at that, was a 500 bed quarantine facility at Hamilton's neighbouring suburb of Pinkenba, being a repurpose of the Commonwealth owned 30-hectare Damascus army barracks.

---

[457] Jonathan Chancellor, 'Riverfront Rush', Mansion Australia, *The Weekend Australian*, 14-15 May 2022, p. 18.

[458] Ibid.

The quarantine station idea was conceived during the Corona Virus crisis when the reflexive and instantaneous policy of hotel quarantine proved inadequate to constrain transmission of an airborne virus. By 2022 however policies of lockdown and suppression were supplanted by the exact reverse: a need to 'live with' the virus, thereby causing the Federal Government to consider alternative uses for Pinkenba. All of which led *The Courier-Mail* to editorialise that Pinkenba, and also the State Government funded alternative quarantine centre at Wellcamp, near Toowoomba, as '*examples of some of the worst pandemic decision-making ... political brinkmanship at its worst, and a waste of taxpayer money*'.[459]

Now would be the perfect opportunity to repurpose wasted taxpayer money to money well spent; all three tiers of government need to swiftly repurpose the Damascus barracks so as to kick-start gentrification of the broadly industrial suburb of Pinkenba into a long term flood-free location for housing. That there is, by '*good luck*',[460] an existing, albeit disused, railway line terminating at Pinkenba means, according to eminent Brisbane architect Ed Haysom, that '*this rail line has potentially enormous benefits if we are looking to increase housing density without increasing the load on the city's roads. The infrastructure is already there to connect whatever is built back to the city* [of Brisbane]'.[461] To house large numbers of persons in a flood-free area with a genuine non-road alternative means of egress and ingress potentially available to be refurbished and reinstated is an idea that is inherently sensible. '*Let us hope someone in Economic Development has Queensland Rail's phone number*',[462] Haysom concluded. Quite so.

Dams will continue to be debated; for the reasons outlined in Chapter 4, and also as a function of the '*one million more*' in population Schrinner predicts, new dams are propounded to be a non-so-

---

459 'Pandemic lessons must be learnt', Editorial, *The Courier-Mail*, 10 June 2022, p. 28.

460 Ed Haysom, 'No rail line to service Olympic athletes' village', *My Village News*, August 2022, p. 15.

461 Ibid.

462 Ibid.

lution. Apart from Linville, far too far north in the Upper Brisbane River catchment to hold back any flood waters of any substance,[463] wherever else closer to Brisbane City where dams may theoretically be viable in an engineering sense will carry an opportunity cost of forever foregone productive farmland and/or land on which to situate housing developments for the '*one million more*'.

Even more so where land that has previously been used for housing substantial numbers of people becomes increasingly untenable: Rocklea and Goodna being prime examples of established suburbs whose futures are anything but guaranteed.

Moreover the definition of 'farming' has also radically changed as times evolve; climate change was cited in Chapter 8, and substantially more during the 2022 federal election campaign. Hence the conversion of farming lands from growing and raising traditional farming outputs like plants and animals to the installation of solar panels and wind turbines. As this inexorable trend increases, necessarily even less land will be able to be repurposed for damming. Or housing for that matter, because '[N]*o matter how you look at it, harnessing energy is a land-use issue. This is true whether we are digging coal … fracking natural gas … mining uranium …* [or] *installing massive wind turbines or giant arrays of solar*'.[464]

Construction of a $2 billion wind farm, Australia's largest, capable of generating 1026 MW of electricity began in June 2002, in South-East Queensland: the Macintyre Wind Precinct, south-east of Warwick.[465] Six months later both the size and level of that particular investment doubled to 2000 MW and $4 billion respectively,[466] expanding its land footprint even further. This is because solar panels and wind turbine towers require vast tracts of land to stage sufficient quantities of arrays to generate electricity at commercially sus-

---

463 Supra, fn 95.

464 Saul Griffith, *The Big Switch: Australia's electric future*, Black Inc, an imprint of Schwartz Books Pty Ltd, 2022, p. 35.

465 Perry Williams, 'Set renewable target high: Acciona', *The Australian*, 2 June 2022, p. 15.

466 'Wind farm to expand', *The Australian*, 29 November 2022, p. 15.

tainable levels; to meet its emission reduction targets, global mining giant Rio Tinto is scoping two separate tracts of no less than 200 hectares each on which to build twin 100-megawatt solar farms in a bid to offset emissions caused by its extensive Western Australian iron ore operations.[467] The reason for the vast size of these installations being the mere 10% of actual conversion of solar radiation into electricity:

> [T]*he average incident solar radiation on the ground is around 300 watts per square metre. A typical high-quality, modern PV* [photovoltaic] *module ... captures one fifth of this energy ... It is difficult to cover 100% of a space with solar because you need walking space between modules for cleaning and so forth, so let's assume 50% ground coverage ... the inverter that turns the captured energy into ... 240 volt electricity ... is about 95% efficient. This means we get, very approximately, 30 watts per square metre of energy on land covered with a solar farm.*[468]

Self evidently, at a rate of 30 watts per square metre, that's a vast amount of land requiring purposing or re-purposing to solar generation so as to generate sufficient quantities of base load power.

All of this analysis proves that the contemporary growth of alternative non-carbon emitting energy infrastructure axiomatically lowers the supplies of land capable of being dammed, or, to begin to rectify the other major crisis bedevilling Queensland, being able to house residents.

'*Building back better*' may be an epithet of the current era, beloved of politicians desperate for a soundbite. And, to be fair, re-building houses with structures impervious to water and/or deregulating zoning restrictions so as to allow as of homeowner right raising of houses to hopefully evade future floods is a pragmatic policy shift that the Brisbane City Council rightly implemented immediately after 2011. The de Jersey report recommendation that the Flood Re-

467 Brad Thompson, 'Rio's new $900m solar farms to spark cut in emissions', *The Australian Financial Review*, 1 December 2022, p. 19.
468 Griffith, op. cit., pp. 40-41.

silient Home Program be continued, and expanded, by identifying other flood affected areas that will benefit from the program and inviting additional homeowners to participate,[469] is a welcome addition to the debate.

The only aspect of the debate which, to this writer's researches, has not been ventilated properly, if at all, by anyone, is a mindset change at population level when it comes to water conservation conceptually. Schrinner's arguments of 10 May 2022 include a '*nightmare*' – his concept – that '*Games organisers are … forced to warn visitors to severely restrict their water use. Some bright spark thought a nice nostalgic touch would be to hand out those egg-timers that were used during the Millennium Drought so people limit their time in the shower*.'[470] I remember those days, and presume that that '*bright spark*' reference was meant to have been to then Premier of Queensland Peter Beattie who conceived of the idea: '*Peter Beattie's egg timers … Rolled out during the Millennium Drought, we were all told to use them so our showers lasted no longer than a miserly four minutes*'.[471] The populace were not particularly enamoured of this policy, so it is conceivable that persons may, at face value, agree with Schrinner's '*nightmare*' epithet if a repeat is invoked, especially in the immediate lead up to the 2032 Games when '[H]*aving so many people visiting Brisbane for that two weeks … is like inviting many people around to dinner so to show your best face you must clean up your place and put out your best cutlery*'.[472]

And hide away your egg timers.

Jokes aside however, and to be frank, compared to the Armageddon scenario of running out of water altogether, shorter showers were minor inconveniences; an egg-timed shower is still

---

469 Hon Paul de Jersey, *Brisbane City Council 2022 Flood Review*, 9 May 2022, p. 58.

470 Adrian Schrinner, 'Drought could kill the Games if we don't act', *The Courier-Mail*, 10 May 2022, p. 33.

471 Steven Wardill, 'Hard-hatted politics fails to plan for the future of Queensland', *The Courier-Mail* [online], 13 April 2018.

472 Ed Haysom, 'Cities must get their infrastructure right for them to grow well', *My Village News*, November 2022, p. 12.

a shower and adequate for purpose. The modern Brisbane City Council therefore needs to lead the way in replacing mindsets where water restricting in domestic contexts becomes, far from '*nightmare*' scenarios, the norm. Schrinner's opinion piece is a genuinely good start, even if, perhaps unintentionally, his call some four months later for a pre-emptive water release out of Wivenhoe Dam's water storage compartment, duly heeded not long after, could in fact part contribute to the very '*nightmare*' he hoped would not beset Brisbane's 2032 Olympic Games. Maybe then the State Government's October 2022 response to Schrinner's call by deliberately dropping Wivenhoe Dam's water storage capacity from 90.5 to 80 percent could be the very catalyst for words to be replaced by actions. Because shorter showers are not a nightmare; a rapidly growing city running out of drinking water is.

Even less worthy a pretext to apply drinking water is gardening and, perish the thought, lawn sprinkling. Bucket only plant-watering in 2007 may not have been particularly pleasant, yet it did the job policy-wise: 15.08% Wivenhoe Dam capacity was, if not the precise nadir, very close to it; rains re-occurred and eventually levels rose and restored. Brisbane was not rendered uninhabitable. The 6 February 2007 invocation by Malcolm Turnbull – then Minister for the Environment and Water Resources – of pensioners who pull muscles or crack backs caused by bucket watering '*their dry and desiccated lawns and their dead roses*', and his associated recommendation that '*they remember the Leader of the Opposition* [Rudd]',[473] did not turn out as his then party would have hoped. If that particular parliamentary statement was an exhortation to that cohort of people to turn their anger upon Kevin Rudd, the person Turnbull blamed for the causation of that scenario allegedly as a result of the permanent cancellation of the Wolffdene Dam proposal,[474] then it fell on deaf ears. Rudd won that poll with a primary vote for the Australian Labor Party of 43.4%,[475] more than 10% more than the 32.6% that

473 *Tennyson Breach*, p. 74.

474 Ibid.

475 https://www.abc.net.au/elections/federal/2007/

party received at the 21 May 2022 poll[476] that reinstated it, for the first time since Rudd's 2007 victory, to government at federal level with a majority on the floor of the House of Representatives.

What does such a strong victory for Rudd in light of the prolonged and personal criticisms of him on the floor of the 2007 federal parliament prove? Apart from the inherently uncertain, shocking even, nature of election campaigns – for example the sudden conversion of a conservative state politically to '*Greensland*' during May 2022, and Aaron Patrick's prediction of Turnbull's '*support for independent candidates challenging the Liberal Party* … [having] *the potential to alter history*'[477] being fulfilled by '*the strong showing of the … teal independents …*',[478] it proves that a willingness by Queenslanders during a year as seriously drought blighted as 2007 to embrace inconvenient policy outcomes exists if the imperative does also. As it demonstrably did so exist. And with one million more people forecast to move to the environs of South-East Queensland between 2022 and 2032, with no guarantee that the outsize rainfalls of 2022 will be repeated, it is untenable to allow lawn and garden watering, let alone on the basis of flawed assumptions that dams will always fill and remain full.

'*It is hard to find any academics or resource managers who believe that demand side management can do anything but reduce load on the* [South-East Queensland water supply infrastructure] *system in peak periods*',[479] opined Bruce Kingston in a study devoted to disabusing the notion of desalination of Gold Coast ocean waters as a viable long-term water supply solution for South-East Queensland. This writer fully agrees with Kingston's chapter thesis, contained,

---

476 https://www.abc.net.au/news/elections/federal/2022/results/party-totals

477 Aaron Patrick, *Ego Malcolm Turnbull and the Liberal Party's Civil War*, Harper Collins, 2022, p. 300.

478 Graham Lloyd, 'Severe weather amped up climate "doomism" vote', *The Australian*, 25 May 2022, p. 13.

479 Bruce Kingston, 'The sad history of water politics in SE Qld and the bad politics that led to the 'solution' of desalination', Case Study 1, *White Elephant Stampede* [Eds: David Gration, Bruce Kingston and Scott Prasser], Connor Court Publishing 2022, p. 35.

most appropriately, in a chronicle of case studies in policy and project management failures. At the same time, this writer will take up Kingston's challenge by placing his own head above the proverbial parapet by publicly advocating for demand management – and aggressive demand management at that – as the only true water policy solution for this region. Mindsets must change: change away from supply side 'solutions' – including the delusion that sufficient vacant lands exist for another Premier of Queensland to embark on a Sir Joh style dam building spree so as to enable him or her to also declare at premiership end '*Clive I've built a lot of dams*' – to whole of community acceptance of the imperative of reducing overall demand. In the same way that more and more houses presently display yellow-liveried corflutes exhorting '*Climate Action Now*', sloganeering '*Water conservation now*' alongside such corflutes ought similarly be demonstrated. Especially as resource conservation generally, conceptually speaking, is entirely consistent and complimentary with the climate change influenced imperative of reducing carbon emissions.

Just as flawed is the parallel assumption to the reflexive cry of 'build more dams': that the dams already in existence will 'save' Brisbane. 2022 proved that 2011 was not an aberration. If anything, the reverse is true so far as total quantum of rainfalls in Brisbane city itself and surrounds were concerned. It would therefore be churlish not to thank those charged with the operation of Wivenhoe Dam during the final days of February 2022, especially the engineers who devised the '*hack*' which, with a mere 39 centimetres in dam wall height in hand,[480] kept Wivenhoe Dam's gates shut so as to have prevented burdening an already sodden city with even more water. The total quantum of Brisbane area rainfalls in 2022 exceeded those of 1974 – 792.8mm vs 655.8mm[481] – yet the resulting substantially lower impact at the City Gauge {3.85m compared to 5.45m}[482] can not be denied.

---

480 Supra, fn 257 and fn 264.

481 Supra, fn 16.

482 Supra, fn 50.

However the fact remains that before the February 2022 'rain bomb' struck, on Thursday 24 February 2022 Wivenhoe '*had approximately 43 percent of its drinking water supply storage available and 100 per cent of its flood mitigation capacity*'.[483] Put another way, Wivenhoe Dam possessed 43% of capacity in the drinking water compartment before any recourse to the flood mitigation compartment became necessary.[484] Wivenhoe Dam would have possessed an even greater 63.7% capacity in the drinking water storage compartment had the immense rains of February 2022 fallen in January 2021 instead due to its worryingly low 36.3% level as of 13 months earlier.[485] Proving the inexorable nexus between Wivenhoe Dam *capacity* and extreme rainfall events, not Wivenhoe Dam itself. The attendant ability of its engineers to have devised what they devised during late February 2022 by withholding all but the most minor of releases until after the rain bomb departed and sunny days returned at the commencement of March 2022 meant that the downstream impacts in Brisbane were not multiplied and magnified in February 2022. Albeit with only 39 centimetres of Wivenhoe Dam wall height in reserve, but a win is, to quote the well-worn phrase, a win. The prolonged drought of 2014 – 31 October 2020 had a silver lining after all.

Even if persons and properties affected regardless from the Brisbane River's independently conveyed flows of floodwater in Goodna, Rocklea, Rosalie and Fish Lane would definitely have failed to have seen it that way.

Imagine instead that there had been average to better than average rainfalls during the period between 2015 and 2022 with consequential full storage of Wivenhoe's water storage compartment instead of the actual 57% as of February 2022? As there had been up to and including 2 January 2011, and the drinking water compartment of Wivenhoe Dam's percentage was closer to 100% according-

---

[483] Lydia Lynch '*Dam overflow stepped up as rains return*' *The Australian*, 3 March 2022, p. 5.

[484] Lydia Lynch, 'Dam overflow stepped up as rains return', *The Australian*, 3 March 2022, p. 5.

[485] Supra, fn 8.

ly? Forcing the height of Wivenhoe Dam rapidly upwards beyond the compulsory full gate release threshold of 75 metres? The resulting loss and damage within the already flood water laden broader city of Brisbane of necessarily much larger quantities of Wivenhoe released waters meeting those independently 'rain bomb' dropped waters would have been exponentially worse, especially when bearing in mind the way in which the current iteration of the Wivenhoe Dam operations Manual almost halves the cubic metre per second threshold amount of waters at Moggill so far as compliance with a criterion of flooding not more than around 100 residential and commercial premises is concerned.[486]

Reinforcing even further that, in a fully built-out, next to no greenfield sites, floodplain city which Brisbane is, and moreover a city fed by many more watercourses than the Brisbane River exclusively, dams are anything but a silver bullet which, if only they could be operated 'properly', would permanently floodproof the city:

> [T]*he ... flood damage experienced ... and the repeating theme of surprise when this occurs, suggests that we may have both overestimated the capability of these devices* [Somerset and Wivenhoe Dams] *to mitigate floods and failed to learn from history.*[487]

So very true.

According to one commentator, opining in the immediate aftermath of the 2011 flood event, the history of which we have failed to learn from dates as far back as Roman times:

> *We are slow learners. Even Roman town planners knew to avoid building on flood plains. They sacrificed a local lamb and examined its liver before starting a building project. If the organ was discoloured, it would suggest the presence of liver flukes, which lived in low-lying areas prone to flooding.'*[488]

---

486 Revision 16 of the Manual, released November 2021, p. 30.

487 Dr Luke Toombes and Rob Ayre, *Resetting Expectations of Brisbane's Flood Protection*, p. 12.

488 Des Houghton, 'Paying for a river view', *The Courier-Mail*, 22–23 January 2011, pp. 62 and 63.

Lambs not having been native to Australia, even if John Oxley in 1823 was himself appraised of this clever, if a little grisly, kernel of Roman wisdom, he could not have performed the same 'experiment' during his reconnaissance of the then unnamed Brisbane River. Had Oxley instead encountered modern day Brisbane during or immediately after the 1841 flood height, it is fair to infer that his recommendation to establish settlement would definitely not have been made. However, and as already explained in this chapter, matters proceeded apace after Oxley's first, and crucially flood-free, impression during November 1823; as is said in the classics, first impressions are lasting ones, and this definitely appears to have been the case when charting the reasoning behind the gestation of modern day Brisbane. Following Oxley's positivity about his discovery some seven months earlier, the 10 June 1824 edition of the *Sydney Gazette* reported the then administration's '*intention ... almost immediately, to form a Settlement at Moreton Bay, or in its vicinity. The Colonists may anticipate much in favor* [sic] *of Australia's prosperity from the establishment of this contemplated Dependency. New and valuable discoveries will inevitably be the consequence of such a measure*'.[489] On 28 August 1824, Governor Brisbane formally wrote to John Oxley to advise '[T]*he Government vessel the Amity, fraught with troops, convicts and stores ... is placed under your orders for the purpose of crowning your late discovery of a large river flowing into Moreton Bay with the formation of a new settlement in its vicinity*'.[490]

Following Oxley's carrying out of those instructions, during November 1824 Governor Brisbane visited in person, for the first time, and endorsed Oxley's original recommendation to establish a settlement at the junction of modern day Breakfast Creek and the Brisbane River,[491] albeit that the actual settlement so created thereafter centred on what is now Brisbane's central business district. Though the reasons for why Oxley and in turn Brisbane were ignored are

[489] J. G. Steele, *Brisbane Town in Convict Days, 1824-1842*, University of Qld Press, 1975, p. 1, mis-placed capitalisation in original.

[490] Ibid., p. 5.

[491] Ibid., p. 21.

unclear, this location was and remains bound on three sides by the Brisbane River and to the north by a ridge – modern day Wickham Terrace and the aptly named suburb Spring Hill {for an elevated hill above the city is what it is}. As such, the convenient hemming in and self-contained character practically suited the initial settlement pretext of a penal colony[492] compared to the much flatter plain-like nature of the topography of Newstead and Albion where Breakfast Creek and the Brisbane River presently adjoin.

After John Oxley had piloted the Amity and offloaded its '*troops, convicts and stores*' during September 1824, rather than returning to Sydney, Oxley, along with botanist Allan Cunningham, instead explored the Brisbane River between 16 and 28 September 1824.[493] This time around, and unlike what Oxley did not sight 10 months earlier, Oxley and Cunningham both recorded evidence of flooding near what is now Kholo Creek, upstream of Ipswich.[494] Yet this particular memorandum did not seem to make it to Governor Brisbane. Rather, two months later, Oxley had re-returned to Sydney in order to helm Governor Brisbane's first voyage to the nascent '*Settlement at Moreton Bay*'.[495] The diligent researches of historians such as J. G. Steele extract much in the way of correspondence between John Oxley and Governor Brisbane, so the absence of such notification is notable.

What could explain the omission? The very long upstream distance between Kholo Creek and the confluence of the Brisbane River and Breakfast Creek, being the penal colony site advocated for by Oxley, could have had a genuine impact on grounds of relevance. Especially as vessels in those days were anything but high speed, making distances seem in that era even longer than they were. Another objective factor is the much greater width – almost double – of the Brisbane River at its meeting point with Breakfast Creek

---

[492] *River*, p. 7.

[493] Ibid., p. 16.

[494] Ibid., p. 6.

[495] Supra, fn 489.

compared to its narrow width as it envelops the CBD. The greater a river's width the greater its capacity to convey flood waters rather than have those waters overtop the banks. It would be open to infer then that whatever Oxley and Cunningham eyewitnessed at Kholo Creek was not replicated 'on the ground' at Breakfast Creek.

Plus, of course, Oxley's first impression of the preceding year, which contained no evidence of flooding, with associated personal endorsement delivered to Governor Brisbane. Which in turn led to the resources that that same Governor reflexively devoted less than one year later, when he tasked Oxley with conveying the Amity and its '*troops, convicts and stores*' northwards. Would it have been politically palatable to Governor Brisbane to have deprived the colony of what the *Sydney Gazette* had gushed only three months earlier of '[N]*ew and valuable discoveries ... inevitably ... the consequence of such a measure*'?

While on the topic of not so much valuable discoveries but valuable dispositions, and appropriately for the bicentennial year of Oxley's first impression of the land on which it sits, a 130 year old riverfront property at New Farm which shares the Amity's name sold during 2023 for $20.5 million, making history as the most expensive residential home ever sold in Brisbane.[496] The Sydney Gazette writers of 1824 were most prophetic.

1825 heralded the arrival of a new explorer, whose name also adorns many a South-East Queensland physical feature: Major Edmund Lockyer. The oft-cited in this book Lockyer Creek and associated Lockyer Valley farming district bear his name. Akin to the extreme rapidity of flood water rises witnessed there during the 2011 and 2022 flood events, during September 1825 Major Lockyer journalled at Fernvale – in close downstream proximity to modern day Lockyer Creek's junction with the Brisbane River – that discoloured water '*at half past 7 ... had risen a foot in less than an hour ... the rapidity of the current increased every hour and the river*

---

[496] Elizabeth Tilley, 'Money in the Banks. Riverfront Home's Sale Record', *The Sunday Mail*, 30 April 2023, p. 2.

had risen upwards of eight feet by eleven o'clock ... we experienced extreme difficulty in getting the boats up from the great strength of the stream ...'[497] Meaning that in two wholly separate voyages, one year apart, genuine evidence existed of flooding having taken place in the Brisbane River's upstream reaches.

Yet again however another set of visual recordings made no practical difference; 1825 instead was the year during which the first official exports – being pine logs – from the then nascent '*Brisbane Town*' began,[498] thereby confirming, for the first of many times, the economic positivity forecast in the previous year by the Sydney Gazette.

When the first genuinely recorded flood event occurred in 1841, the combination of 17 years of formalised living and industry in the meantime between 1824 and 1841, the conversion in or about 1839 of the then penal colony into a brand new free-settlement and perhaps deliberately misleading suppression of the news of the 1841 flood meant that the proverbial die had long since been cast: abandonment did not occur; to the contrary, '*Brisbane Town*' converted to the ever-growing capital city of the thereafter promulgated State of Queensland.

Returning to the liver flukes lamb sacrifice metaphor, the precise date or dates on which introduction of lambs from Sydney took place is beyond the scope of this analysis. Suffice to say however, if any as introduced lambs similarly came to demonstrate discoloured livers, by then it was far too late to have taken the Roman option by total withdrawal from the floodplain.

This phenomenon has been described as '*sunk-costs syndrome*': '*... too many projects suffer from the classic 'sunk-costs' syndrome, whereby "poor choices will be persisted with, rather than abandoned" because of the level of resources and political reputations invested in a particular project* [, and as]... *a matter of saving face and political*

[497] J. G. Steele, *The Explorers of the Moreton Bay District*, University of Qld Press, 1972, pp. 179, 195 & 196.

[498] Supra, fn 430.

reputation'.[499] The white-elephant projects compendium Scott Prasser co-edited cited as examples the Gold Coast Desalination Plant, the Murray-Darling Basin water reforms, local council mergers, the COVIDSafe App, the Attack Class Submarines program and, perhaps most bizarrely of all, the truly infamous $3.79 million of Federal Government money spent on a '*Milkshake Video*' containing vision of a female throwing the contents of her milkshake in her male friend's face as a supposed way to teach children about consensual relationships.[500] It, at least, was discontinued.

Classing all of the foregoing as '*white elephant projects*',[501] Prasser distinguished other '*projects which have gone astray because of unpredictable circumstances such as a natural disaster* [which] *are not characterised as a white elephant project unless there was failure to consider, as a normal part of any proper risk analysis ... factors that contributed to the subsequent damage or loss*'.[502]

The township then city of Brisbane was not a mere '*project*', in the way of the above examples. Even if, for the sake of argument, it was, by Prasser's instantly cited criterion Brisbane is not a white elephant, nor could be construed as one, given that there was no possible flood risk analysis that could have been meaningfully performed between 1823 and 1841; the evidence of flooding Oxley and Lockyer had separately noted during 1824 and 1825 was, as explained above, quite some distance upstream from the locations, both originally intended and actual, of the as constructed original penal colony and associated settlement.

However the concept of sunk-costs syndrome, combined with the perhaps deliberately misleading suppression of recording into the 1841 flood, more than adequately describe the ways in which modern day Brisbane's gestation began absent any flood narratives

---

499 Scott Prasser, 'Lessons from the stampede', *White Elephant Stampede* [Eds: David Gration, Bruce Kingston and Scott Prasser], Connor Court Publishing 2022, p. 215.

500 Steven Schwartz AM, *Discontent over Consent: The Short Unhappy Life of the Milkshake Video*, Ibid, pp. 231 and 235.

501 Supra fn 499, p. 214.

502 Supra fn 499, p. 216.

at all, until 1841 when it was far too late to terminate. Instead what arose, reactively, was an 'ignore it and supress it and it will go away' mindset, then, when 1893 abruptly disabused any notion – real or contrived – that Brisbane was not built on a vulnerable floodplain, river damming became the next – and convenient – replacement paradigm. Convenient because it could enable then governments to have answered community expectations that governments in western democracies are, and always have been, expected to '*do something*' about almost every issue because governments can '*fix*' everything[503]. Including, apparently, a river system as vast and diverse as the Brisbane River system is.

Time and time again however, the Brisbane River has proven otherwise.

And it will inevitably continue to do so.

This writer's post-2011 flood event analysis book *Tennyson Breach* was so named to describe the breached contracts, breached banks of the Brisbane River and breached property value and lifestyle dreams of those who bought into the Tennyson Reach development. The 2022 flood event, being so much broader in effect than the much more confined nature of 2011's, especially as a result of the many alternative water sources which themselves contributed to the flooding thanks to the supercharged feeding impact of the 'rain bomb', informs the title to this book. Brisbane was most certainly breached during 2022, much more so than in 2011, despite the lower official flood height. Brisbane will be breached again, if not by flooding, by drought and all the existential fear and distress which droughts bring. The ball is firmly therefore in the court of all three levels of government to render all previous assumptions {or hopes} redundant and grasp the many policy nettles which arise from living in a drought prone city built, literally, on a flood plain.

---

[503] Supra fn 499, p. 219.

Acknowledgements and Copyright

The author would like to thank the following for making this book possible:

My daughter Stephanie Topp for her excellent skills wielding her father's iPhone to take many of the Wivenhoe Dam and associated photos contained herein; Steph shows just as much aptitude with my credit card at shopping centres, so it's good to see her skill-set turn more profitable for once!

To my other photo suppliers Lindsay Millard, Justin Fleming, Andrew Clarkson, Alan Sagan and many more besides; your images greatly enhanced my narrative;

Lawyers are traditionally bereft of any and all knowledge and ability in matters of atmospheric science, maths and physics, so I am greatly indebted to engineers Rob Ayre, Lindsay Millard and Steven Voss and meteorologist Anthony Cornelius for your images, graphs and generalised assistance and patience in receiving numerous 'what about' emails and follow ups from me in your in-box;

The Bureau of Meteorology and Seqwater for your permissions to quote and adapt graphs and materials, again absent which many of the arguments and conclusions contained would not have been possible to have particularised;

Finally to my publisher for believing in the project, Scott Prasser as editorial supervisor and Tim O'Dwyer, Peter Spearitt and Rob Ayre for your cover endorsements.

Index

www.ingramcontent.com/pod-product-compliance
Lightning Source LLC
LaVergne TN
LVHW020054110826
845155LV00022B/81

* 9 7 8 1 9 2 2 8 1 5 7 8 1 *